Why Don't Penguins' Feet Freeze?

And 114 Other Questions

NewScientist

Free Press
New York London Toronto Sydney

*f*P
FREE PRESS
A Division of Simon & Schuster, Inc.
1230 Avenue of the Americas
New York, NY 10020

First Free Press paperback edition June 2007
Originally published in Great Britain in 2006 by Profile Books Ltd.
Published by arrangement with Profile Books Ltd.

FREE PRESS and colophon are
trademarks of Simon & Schuster, Inc.

For information about special discounts for bulk purchases,
please contact Simon & Schuster Special Sales:
1-800-456-6798 or business@simonandschuster.com

Manufactured in the United States of America

10   9   8   7   6

Library of Congress Cataloging-in-Publication Data
Why don't penguins' feet freeze?: and 114 other questions /
New scientist.
       p. cm.
Includes index.
1. Science—Miscellanea. I. Related title: New scientist (1971). II. Title.
Q173 .W625   2007
500—dc22   2006053506

ISBN-13: 978-1-4165-4146-2
ISBN-10:   1-4165-4146-2

_f_**P**

Also by *New Scientist*

Does Anything Eat Wasps?
And 101 Other Unsettling, Witty Answers to Questions
You Never Thought You Wanted to Ask

Stroke a Martian:
And 99 Other Things to Do Before You Die

100 Things to Do Before You Die
(Plus a Few to Do Afterwards)

Contents

Why Don't
Penguins'
Feet Freeze?

Introduction

This book's predecessor—*Does Anything Eat Wasps?*—was the surprise publishing phenomenon of the 2005 holiday season. The quirky collection of science questions and answers from *New Scientist*'s Last Word column took the bestseller lists by storm, leaving those associated with the column's 13-year history a little shell-shocked and breathless. This sense of surprise was reinforced by the fact that *Does Anything Eat Wasps?* was actually the third collection of Last Word questions and answers in book form. The first two had modestly paid their way without ever troubling the best-seller charts. This, on reflection, seems a pity, for those first two volumes contain some of the questions that have come to define exactly what The Last Word represents: the pursuit of the offbeat and the trivial. Why is snot green? Why does grilled cheese go stringy? Why does tinfoil make tooth fillings painful? And, of course, why don't penguins' feet freeze?

Perhaps more important, those first two books also contain answers to questions that are asked every week by readers who newly discover The Last Word. It seems everybody wants to know why the sky is blue or hair turns gray. And you can find the answers on pages 137 and 3, respectively.

Interestingly, the most popular question when the first two Last Word books were translated into German was "Why don't sleeping birds fall out of trees?" This led to the longest title of any *New Scientist* book published anywhere in the world—*Warum fallen schlafende Vögel nicht vom Baum?* And although the title *Why Don't Penguins' Feet*

Freeze? does not match this for length, this book is by far the thickest and best-value collection of Last Word questions yet published. Since we feel the first two books deserved a wider audience, we've combined the best questions and answers from those volumes with some wholly new material from the weekly magazine column to create this bumper edition. We hope they will keep you entertained for weeks to come.

Does Anything Eat Wasps? generated a vast amount of media coverage, during which I was constantly asked why my book had sold so well. The truth was, of course, that it wasn't my book at all; it belonged to the readers of *New Scientist*. Remember, everything you see here is provided by contributors to The Last Word, both in the weekly print edition of *New Scientist* and online. Thousands of questions are posed every year and just as many answers are received. So if you have something to ask *New Scientist*'s readers visit http://www.newscientist.com/lastword.ns or buy the weekly magazine. Even better, if your friends routinely describe you as a complete know-it-all—or, like me, the barroom bore— you are just the kind of person we are looking for. The Last Word is your natural home, so why not help us answer our endless supply of questions? Without readers' input The Last Word would not survive and, as you'll read here, none of us would know how to toughen up our conkers.

Enjoy this compilation and keep those questions flooding in.

MICK O'HARE

Again, special thanks are due to Jeremy Webb, Lucy Middleton, Alun Anderson, the production and copyediting teams of *New Scientist,* and the people at Profile Books for making this book far better than it might have been. And to Thomas O'Hare for providing an endless source of kiddie questions . . .

1   Our Bodies

 ## Gray Matters

Why does hair turn gray?

KEREN BAGON

Gray (or white) is merely the base "color" of hair. Pigment cells located at the base of each hair follicle produce the natural dominant color of our youth. However, as a person grows older and reaches middle age, more and more of these pigment cells die and color is lost from individual hairs. The result is that a person's hair gradually begins to show more and more gray.

The whole process may take between 10 and 20 years—rarely does a person's entire collection of individual hairs (which, depending on hair loss, can number in the hundreds of thousands) go gray overnight. Interestingly, the color-enhancing cells often speed up pigment production as we age, so hair sometimes darkens temporarily before the pigment cells die.

BOB BARNHURST

⚙ Light Sneeze

I have noticed that many people tend to sneeze when they go from dark conditions into very bright light. What is the reason for this?

D. BOOTHROYD

Photons get up your nose!

STEVE JOSEPH

I think that the answer may be fairly simple: when the sun hits a given area, particularly one shielded or enclosed in glass, there is a marked rise in local temperature. This results in a warming of and subsequent upward movement of the air and, with it, many millions of particles of dust and hair fibers. These particles quite literally get up one's nose within seconds of being elevated—hence the sneezing.

ALAN BESWICK

My mother, one of my sisters, and I all experience this. I feel the behavior is genetic and confers an unrecognized evolutionary advantage. I have questioned many people, and we sun-sneezers seem to be in the minority. However, as the ozone thins and more ultraviolet light penetrates the Earth's atmosphere, it will become increasingly dangerous to allow direct sunlight into the eye. Those of us with the sun-sneeze gene will not be exposed to this, because our eyes automatically close as we sneeze! The rest of the population will gradually go blind, something not usually favored by natural selection.

ALEX HALLATT

The tendency to sneeze on exposure to bright light is termed "the photic sneeze." It is a genetic characteristic, which is transmitted from one generation to the next and which affects between 18 and 35 percent of the population. The sneeze occurs because the protective reflexes of the eyes (in this case on encountering bright light) and nose are closely linked. Likewise, when we sneeze, our eyes close and also water. The photic sneeze is well known as a hazard to pilots of combat planes, especially when they turn toward the sun or are exposed to flares from antiaircraft fire at night.

R. ECCLES
COMMON COLD AND NASAL RESEARCH CENTRE, CARDIFF, UK

Here are some early thoughts on the subject of light sneezing from Francis Bacon's *Sylva Sylvarum* (London: John Haviland for William Lee, 1635, page 170): "Looking against the Sunne, doth induce Sneezing. The Cause is, not the Heating of the Nosthrils; For then the Holding up of the Nostrills against the Sunne, though one Winke, would doe it; But the Drawing downe of the Moisture of the Braine. For it will make the Eyes run with Water; And the Drawing of Moisture to the Eyes, doth draw it to the Nosthrills, by Motion of Consent; And so followeth Sneezing; As contrariwise, the Tickling of the Nosthrills within, doth draw the Moisture to the Nosthrills, and to the Eyes by Consent; For they also will Water. But yet, it hath been observed, that if one be about to Sneeze, the Rubbing of the Eyes, till they run with Water, will prevent it. Whereof the Cause is, for that the Humour, which was descending to the Nosthrills, is diverted to the Eyes."

C. W. HART
SMITHSONIAN INSTITUTION, WASHINGTON, D.C.

 ## Comes in Handy

Why do we have fingerprints? What beneficial purpose could they have evolved to serve?

MARY NEWSHAM

Fingerprints help us in gripping and handling objects in a variety of conditions. They work on the same principle as the tires of a car. Although smooth surfaces are fine for gripping in a dry environment, they are useless in a wet one. So we have evolved a system of troughs and ridges to help channel the water away from the fingertips, leaving a dry surface, which allows a better grip. The unique pattern is merely a phenomenon that is used by the police to identify individuals.

JAMES CURTIS

Fingerprints are the visible parts of rete ridges, where the epidermis of the skin dips down into the dermis, forming an interlocking structure (similar to interlaced fingers). These protect against shearing (sideways) stress, which would otherwise separate the two layers of skin and allow fluid to accumulate in the space (a blister). They appear on skin surfaces that are subject to constant shearing stress, such as fingers, palms, toes, and heels. The unique patterns are simply due to the semi-random way in which the ridges and the structures in the dermis grow.

KEITH LAWRENCE

✿ Crinkle Tips

Why does skin—especially on the fingers and toes—become wrinkled after prolonged immersion in water?

LLOYD UNVERFIRTH

The tips of fingers and toes are covered by a tough, thick layer of skin, which, when soaked for a prolonged period, absorbs water and expands. However, there is no room for this expansion on fingers and toes, so the skin buckles.

STEVEN FRITH

Your whole body does not become crinkled, as the skin has a layer of waterproof keratin on the surface, preventing both water loss and uptake. On the hands and feet, especially at the toes and fingers, this layer of keratin is continually worn away by friction. Water can then penetrate these cells by osmosis and cause them to become turgid.

ROBERT HARRISON

❖ Take the Pils

Why is it that when I walk home from the bar after a few beers, I always stumble to the left more than to the right?

CHRIS WOOD

A similar situation arises when people wander in the forest or desert. They may intend to walk in a straight line; but if they are lost and have no landmarks to guide them, most people will unconsciously walk slightly toward the left, making a big counterclockwise circle, which brings them back to their starting point.

The reason for this is that most people have a slightly stronger, more flexible right leg. This is common knowledge among sports scientists, and most people who have undergone strength tests in their legs can confirm it.

Most people also find they can lift their right leg slightly higher than their left. The right leg has a stronger stride than the left one and so when there are no guiding landmarks a circular walking route is the result.

Also, the slightly greater strength of the right leg means that when you push on the ground with your right foot, the push to the left is slightly greater than the push to the right produced by the left foot. The longer stride and greater push combine to cause most people to move counterclockwise in the course of a long walk.

HAN YING LOKE

The human body is never perfectly symmetrical. In this case, the right leg seems to be longer than the left. A cardboard coaster placed in the left shoe underneath the foot should remedy the problem quite easily.

J. JAMIESON

We each have a dominant eye, which we rely on more than the other, weaker eye. Instinctively, we try to walk where we can see best (although we normally correct this to allow ourselves to walk forward). So when we stumble, it is more likely that we will stumble in the direction of our dominant eye.

This is because the brain, in trying to correct the situation, has to react fast and gives more weight to the information coming from the dominant eye to work out where to put the feet in order to regain balance. Hence the feet tend to be aimed at a position toward the side of your body on which the dominant eye lies, resulting in a stumble in that direction. In this case the questioner's dominant eye is obviously his left.

This phenomenon can be used to steer riding animals— simply cover up one of their eyes and they will tend to move in the direction of their remaining eye.

ADRIAN BAUGH

The questioner obviously walks to the bar with his change in his right pocket and his keys in his left. After he has spent all his money on beer, the weight of his keys pulls him to the left as he walks home.

SIMON THORN

Members of the Department of Physics at Auckland University have held consultations regarding this issue, and our most popular theory derives from an application of the simple principles of gravity gathered from our common experience in returning from pubs in Auckland.

Currency in denominations lower than NZ$10 is mostly in coins, some of them quite large in size. During an evening in the pub, the drinker accumulates a large number of such coins in his or her pocket. Assuming that coinage is similar in your questioner's country and that the questioner habitu-

ally carries his coins in his left pocket, elementary laws of
gravity dictate that his gait will incline to the left. It is not
uncommon for some New Zealanders in similar circumstances to actually walk in a circle.

NELSON CHRISTENSON
UNIVERSITY OF AUCKLAND, NEW ZEALAND

After you have stood for endless hours in a bar with your
beer glass in your right hand, it is inevitable that you are still
subconsciously counterbalancing the glass's weight, and thus
stumbling more to the left. The opposite can be demonstrated in left-handed beer drinkers.

BY E-MAIL, NO NAME SUPPLIED

◈ By the Left

*Why is it that when two people walk together they often
subconsciously start to walk in a synchronized manner? Is
this some natural instinct?*

SIMON APPERLEY

The zoologist and specialist in human behavior Desmond
Morris says that the reason people start to walk like each
other is a subconscious need to show that they agree with,
and so fit in with, their companion. This is also a signal to
other people that "we are together, we are synchronized."

Other studies suggest that we adopt the mannerisms of
our companions as well, especially our superiors, as when we
cross our legs in the same directions as others. An example
often given is that when, in a meeting the boss scratches his
nose, others at the table then follow him without realizing it.

ADITHI

While it is a purely unsubstantiated opinion, I do have an
answer to why people tend to synchronize their steps. Ob-

serving a group of children walking in a park recently, supervised by two adults, I noted that the adults synchronized their steps and direction, while the children walked, ran, and skipped apparently at random, running ahead, lagging behind, and deviating from the common course.

Perhaps these children, unpolluted by society's emphasis on conformity, have not yet learned that it is unacceptable to march to your own drum.

TODD COLLINS

The next time you walk alongside somebody, walk out of step. Then try to follow the conversation you are having. You will soon fall back into step, because once you are in step with the other person, it is easier to watch where you are walking and then turn to look at him or her.

Communication is easier with another person when you are in close proximity and when both faces are relatively stable and not bobbing all over the place.

HAMISH

Here is a more prosaic (less sociologically inclined) explanation. When people walk they have a slight side-to-side sway. Two people walking together and out of step would bump shoulders every second step.

PETER VERSTAPPEN

Helpless Laughter

Why is it that if you tickle yourself it doesn't tickle, but if someone else tickles you, you cannot stand it?

DANIEL (AGE 7) AND NICOLAS (AGE 9) TAKKEN

If someone was tickling you and you managed to remain relaxed, it would not affect you at all. Of course, it would be difficult to stay relaxed, because tickling causes tension for most of us, such as feelings of unease due to physical contact, the lack of control, and the fear of whether it will tickle or hurt. However, some people are not ticklish—those who for some reason do not get tense.

When you try to tickle yourself, you are in complete control of the situation. There is no need to get tense and therefore no reaction. You will notice the same effect if you close your eyes, breathe calmly, and manage to relax the next time someone tickles you.

The laughter is the result of the mild state of panic you are in. This may be inconsistent with "survival of the fittest" theories, because panic makes you more vulnerable. But as in many cases, nature is not necessarily logical.

SIGURD HERMANSSON

Live Wire

Where does the force come from when you are thrown horizontally across a room after touching a live electrical connection? I thought there was a reaction to every action, but there is no obvious push from the electricity.

JOHN DAVIES

The force comes from your own muscles. When a large electrical current runs through your body, your muscles are stimulated to contract powerfully—often much harder than they can be made to contract voluntarily.

Normally the body sets limits on the proportion of muscle fibers that can voluntarily contract at once. Extreme stress can cause the body to raise these limits, allowing greater exertion at the cost of possible injury. This is the basis of the "hysterical strength" effect that, famously, allows mothers to lift cars if their child is trapped underneath, or gives a psychotic the strength to overcome several nursing attendants.

When muscles are stimulated by an electric current, these built-in limits don't apply, so the contractions can be violent. The electric current typically flows into one arm, through the abdomen, and out of one or both legs; this process can cause most of the muscles in the body to contract at once. The results are unpredictable, but given the strength of the leg and back muscles, can often send victims flying across the room with no voluntary action on their part. When this effect is combined with an unexpected electric shock, you feel as if you are being flung, rather than flinging yourself.

The distance people can involuntarily fling themselves can be astonishing. In one case a woman in a wet parking lot was hit by lightning. When she recovered she found herself some 35 feet from where she had been struck. However, in this case there may also have been some physical force involved from a steam explosion, as water on her and the area in which she was standing was flash-boiled by the lightning. She survived, though she was partially disabled by nerve damage and other injuries.

A common side effect of being thrown across the room by an electric shock, apart from bruising and other injuries, is muscle sprain caused by the extreme muscle contractions. These contractions can also damage joint and connective tissue. Physiotherapists, chiropractors, and osteopaths might consider asking new patients if they have ever experienced an electric shock.

Being thrown across the room can save your life by breaking the electrical contact. In other cases, particularly when the source of the current is something they are holding, the victims' arms and hand muscles may lock on to it. They are unable to let go and, if nothing else intervenes, they may die from heart fibrillation or electrocution.

I recall what may be an apocryphal account of a poorly grounded metal microphone causing a rock singer to be involuntarily locked onto it. Unfortunately, writhing on the floor while screaming incoherently was not entirely unusual during his shows, and it was a while before one of his road crew figured out that something was amiss and killed the power.

ROGER DEARNALEY

It is interesting to consider why the subject is thrown across the room rather than freezing in a tetanic posture. This is because some muscle groups dominate others. Compare this with the muscle effects seen in a stroke victim: if the stroke is severe enough so that no cerebral control is present over one side of the body, the arm is held flexed (that is, wrist bent with fingers pointing to the wrist, elbow bent so that the forearm meets the upper arm) and the leg extended (knee straight, ankle extended so that the toes point to the ground).

This is because without cerebral control, the spinal cord reflexes cause all muscle groups to be active, including both components of any bending and straightening muscle pairs. The dominance of one muscle group over another produces the effect described.

Therefore, if any electrical charge triggers all muscle groups, the imbalance in "bending and straightening" muscle pairs produces the force that is required to throw the person across the room.

It's not at all recommended, but I have heard that if you touch a conductor carrying a current using the back of your hand it is safer than the palm because the resultant muscle spasm does not force you to grip the conductor, producing a continuing shock.

There is always the effect on the heart to consider, too, but that is another matter.

JOHN PARRY

Left in Doubt

As a left-handed person I was both amused and annoyed by an article in New Scientist *suggesting that left-handed people are at greater risk of accidental death. How can this be? Surely a right-handed person has just as much change of dying accidentally as I do. Or is there some unknown factor involved?*

ALAN PARKER

Approaching obstacles, right-handed people will, in general, circumvent them by going to the right, while left-handed people will go to the left. If two same-handed people approach an obstacle from the opposite direction they will walk safely around it without bumping into one another en route. If two people of different handedness approach an obstacle from the opposite direction, they will pass on the same side and, potentially, may bump. Because most people are right-handed, it is left-handed people who most frequently find themselves bumped in these situations. This is a simple example, but if it is taken to extremes and multiplied by a lifetime of bumps, the result is a shorter life expectancy for left-handed people.

HANNAH BEN-ZVI

We left-handers are at greater risk of accidental death because industrial tools and machinery are designed for the right-handed. Left-handers are, therefore, more likely to chop off parts of themselves in all manner of mechanical devices.

An interesting example is the SA-80 assault rifle. When fired from the left shoulder, it ejects spent cartridges, at great velocity, into the user's right eye.

DANIEL BRISTOW

✦ No Flakes

How does antidandruff shampoo work?

EUGENE

Dandruff is thought to be caused by overgrowth of yeasts such as *Pityrosporum ovale* which live on normal skin. This overgrowth causes local irritation resulting in hyperprolifer-ation of the cells (keratinocytes) which form the outer layer of the skin. These form scales which accumulate and are shed as dandruff flakes.

Antidandruff shampoos work by three mechanisms: in-gredients such as coal tar are antikeratostatic and they in-hibit keratinocyte cell division; detergents in the shampoo are keratolytic, breaking up accumulations of scale; finally, antifungal agents such as ketoconazole inhibit growth of the yeast itself. Other components such as selenium sulfide also inhibit yeast growth and therefore inhibit scaling.

RODDIE MCKENZIE
UNIVERSITY OF EDINBURGH, UK

❖ Gas Gassing

Why does speaking through helium raise the frequency of the sounds emitted, even when the final transmission to the hearer is through air?

DAVID BOLTON

Sound travels faster in helium than in air because helium atoms (atomic mass 4) are lighter than nitrogen and oxy-gen molecules (molecular mass 14 and 16 respectively). In the voice, as in all wind instruments, the sound is produced as a standing wave in a column of gas, normally air. A sound wave's frequency multiplied by its wavelength is equal to the speed of sound. The wavelength is fixed by the shape of the mouth, nose, and throat; so, if the speed of sound increases, the frequency must do the same. Once sound leaves the mouth its frequency is fixed, so the sound arrives to you at the same pitch as it left the speaker. Imag-ine a roller-coaster ride. The car speeds up and slows down as it goes around the track, but all cars follow exactly the same pattern. If one sets out every 30 seconds, all will reach the end at the same rate, whatever happens in between.

In stringed instruments, the pitch depends on the length, thickness, and tension of the string, so the instrument is unaffected by the composition of the air. Releasing helium in the middle of an orchestra would therefore create havoc. The wind and brass would rise in pitch, while the pitch of the strings and percussion would remain more or less the same. In the *Song of the White Horse* by David Belford, the lead soprano is required to breathe in helium to reach the extremely high top note.

EOIN MCAULEY

❖ Brain Waves

Why are there fissures or folds in the surface of the brain?

BRIAN LASSEN

The brain has fissures to increase the surface area for the cortex. Dimmer animals such as rats have smooth brains.

Much of the work carried out in the brain is performed by the top few layers of cells—a lot of the brain's volume is, in effect, point-to-point wiring.

So, if you need to do lots of processing, it is much more efficient to grow fissures than it is to expand the surface area of the brain by increasing the skull diameter.

ANTHONY STAINES

Evidently they are there to maximize the surface area of the brain cortex. The real question is why this is necessary. The answer probably lies in the relative number of short-range and long-range connections needed.

If many short-range connections are required, it makes more sense to pack the processing units into thin, almost two-dimensional, plates and reserve a third dimension for long-range connections.

If the neurons were distributed homogeneously over the whole volume of the brain, long-range connections would possibly be shorter, but they would take up the space between the computational units of the brain and thus lengthen the short-range connections, increasing the overall brain volume.

JANNE SINKKONEN

Another possible answer lies in the amount of heat produced in the brain—Ed.

Brain tissues consume very large amounts of energy, and the resulting heat that is generated has to be dumped. Put your hand on your head and notice how hot it feels compared with your thigh.

Brains of lower vertebrate animals lack extensive folds because they have relatively less heat to get rid of.

Humans, on the other hand, have large brains which do a lot of work. The extra folds in our brains increase the surface area for blood vessels to dump the excess heat produced by all that hard thinking. If our brains were to evolve into more complex and larger organs, their folding would have to increase exponentially in order to be able to release the additional heat that they would produce.

GERALD LEGG

Many intelligent vertebrates are endowed with both large brains and a very convoluted cerebral cortex. Therefore, although the dolphin and the shark are of similar size, the dolphin's brain is considerably larger and more convoluted than the shark's.

The cat and the rabbit are also of similar size, but the cat, being carnivorous, has a more complex lifestyle, presumably necessitating greater intelligence, so the cat has a convoluted brain while the rabbit does not.

The size of the animal is also an important factor. Mice and rats, while showing intelligent behavior, have hardly any fissures in their brains, but elephants and whales have brains that are even more convoluted than a human's.

It is interesting that this larger amount of cerebral cortex does not necessarily correspond to a larger number of cortical nerve cells. It turns out that these are larger and more widely spaced in large animals.

One reason for this is that the ratio of glia to neurons is considerably greater in these large vertebrates. As a result, the cerebral cortex—a laminar structure—needs to become folded to contain the number of neurons that smaller animals can afford to have in a nonfolded cortex.

E. RAMON MOLINER

◈ Concentration

People doing a tricky job will stick their tongue out and clamp it between their lips. Why? Does this happen in all cultures?

STEVE TOWNSEND

When you need to concentrate on something—say, a word problem—you are using the hemisphere of the brain also used for processing motor input. It is amusing to see people slow down when they are thinking of something difficult while walking. This is caused by interference from the two activities fighting for the same bit of brain to process them. I suppose by biting your lip or sticking your tongue out, you are suspending motor activity and also keeping your head rigid to minimize movement and interference.

MELANIE WESTERN

Large areas of the brain are devoted to control of the tongue and to the receipt of sensation from it.

Perhaps if the tongue is held rigid against the teeth or lips, the activity of those areas is subdued, allowing delicate tasks like threading a needle to proceed with less interference.

BARRY LORD

⚙ What's the Crack?

What causes the noise when you crack your knuckles or any other joint?

MARTY BROWN

A click or crack is often heard when a joint is moved or stretched. When the pressure of the synovial fluid in the joint cavity is reduced, this may create a gas bubble and generate a popping sound. The sound may also be the result of separating the joint's surfaces, which releases the vacuum seal of the joint.

These noises are sometimes produced during osteopathic treatment, but this does not prove that the treatment has worked; nor does their absence mean the treatment has failed. The test of success is whether the joint's range and ease of movement have been improved.

WILL PODMORE
BRITISH SCHOOL OF OSTEOPATHY
LONDON, UK

All the soft tissues of the body, including the capsules of joints, contain dissolved nitrogen. When a vacuum is applied to the joint space by pulling on the bones—say, by flexing the fingers strongly—nitrogen suddenly comes out of solution and enters the joint space with a slight popping sound.

Radiologists often see a small crescent of gas between the cartilages of the shoulder joint on the chest X-rays of children who are held by the arms. This is due to the force of pulling on the arms, which causes nitrogen to evaporate into the joint space. It can infrequently be seen in the hip, too.

Small, highly mobile bubbles sometimes appear within the hip joint of a baby being screened by ultrasound for congenital dislocation of the hip. This usually happens if the infant is struggling and has to be held firmly. The bubbles disappear after a short while when the nitrogen dissolves again.

If the fingers were X-rayed immediately after the knuckles were cracked, a fine lucency—a result of thousands of tiny opaque bubbles—would probably be visible between the ends of the bones.

TONY LAMONT
MATER CHILDREN'S HOSPITAL
BRISBANE, QUEENSLAND, AUSTRALIA

Wine into Water

No matter what color drink one consumes, when the liquid finally leaves the body the color has gone. What happens to it?

P. BEEHAM

The liquid that leaves the body is almost unrelated, in chemical composition, to the liquid consumed. Any substance, solid or liquid, that goes down the esophagus passes through the digestive tract and, if not absorbed, is incorporated into the fecal matter. Urine, in contrast, is created by the kidneys from metabolic waste produced in the tissues and transported through the bloodstream.

Any colored compound that you drink either will or will not interact biochemically with the body's systems. If it does interact, this interaction (like any other chemical reaction it might undergo) will tend to alter or eliminate its color. If it does not interact, the digestive system will usually decline to absorb it and it will be excreted in the feces—which, you will have noticed, show considerably more color variation than the urine.

STEPHEN GISSELBRECHT

Colored substances in food and drink are usually organic compounds that the human body has an amazing ability to metabolize, turning them into colorless carbon dioxide, water, and urea. The toughest stuff is often taken care of by the liver, which is a veritable waste incinerator. However, on the very infrequent occasions when the intake of colored substances exceeds what the body can quickly metabolize, the color is not necessarily removed as the liquid leaves the body. This is well known to anyone who has indulged in large quantities of borscht (Russian beet soup).

HANS STARNBERG

 ## Grave Concern

A friend's grandfather was exhumed a little while ago in southern Italy in order to be reburied next to his recently deceased wife. Amazingly, his body was found to be completely intact and no decomposition at all seemed to have taken place. Yet he died about 30 years ago from his injuries in a serious car accident and had been buried in an ordinary coffin. How can a body not decompose in this time? Is this a common occurrence? Is soil or geography important?

KIRA KAY

Non-decay of a dead body is more common than most people suppose. Many saints have had their claim to sainthood upheld by the nifty trick of not putrefying after burial. More ordinary examples include the wife of Dante Gabriel Rossetti, who troubled him somewhat by being revealed in all her undecayed glory when, short of money and lacking fresh inspiration, he broke into her grave to steal back the poems he had buried with her.

This type of preservation happens when the adipose tissue in the body forms adipocere, a soapy-textured substance, composed mainly of saturated fatty acids and salts of fatty acids. The colloquial term for an adipocere-ridden corpse is a "soap mummy."

Women tend to be preserved more often than men, probably because they have more fat to start with, and conditions such as humidity and warmth also have an effect. The dead person described in the question, having been buried in southern Italy, probably had a better chance of preservation than he would have had if he had been stuck in the cold mud of England; some very well-preserved adipocere corpses have been discovered in Italy.

Adipocere can either form quickly, within weeks, or after several years. In the latter case a body may reach quite an advanced stage of decay before the development of adipocere sets in. It helps if a body is overweight, as an obese corpse contains enough water and fat to start adipocere formation quickly, regardless of the burial conditions. Adipocere can also be encouraged by covering the body in clothing or a shroud made of artificial fibers, by moist conditions, and by the presence of a substance such as formaldehyde. In rare cases, not only fat but also muscle turns to adipocere. If the body was in very good condition, this might have been the case.

ANNE ROONEY

For a body to putrefy in a grave, there needs to be enough moisture to allow tissue breakdown both from autolysis and from the action of microorganisms, usually starting in the ileocecal region of the intestine. In arid conditions, including dry soil, the corpse will lose water, principally by evaporation as the drier material around it draws the water away. This could even occur through the walls of a wooden coffin, provided the surrounding soil was dry enough to continue absorbing water and conditions were warm enough to encourage evaporation.

The location of the grave—southern Italy—suggests that these conditions were present, and this is probably what stopped putrefaction. Indeed, bodies left aboveground can be partially preserved by this process—for example, in haylofts, where the surrounding dry hay and air draw water out of the dead body.

An extension of this process is found in natural graves in arid regions that have correspondingly dry soil, to the point at which nearly all the water is removed, leaving dry, leathery tissues. This is mummification, and its natural occurrence in the dry sands of ancient Egypt probably encouraged mummification there as a cultural practice.

ALAN TAMAN

 # That's Life

What chemical formula would accurately describe an adult human being, in terms of the relative distribution of elements (including pollutants)? And what might be the formula for the first alien life-form we encounter?

PAUL MONTMORENCY

One's "chemical formula" depends on a number of factors, most notably whether we're talking about a he or a she. Male bodies contain more water than female bodies, which have extra lipids. By weight, oxygen amounts to about two-thirds of the body, followed by carbon at 20 percent, hydrogen at 10 percent, and nitrogen at 3 percent. Elements originating from pollutants would be present in only trace amounts.

If a human body were broken into single atoms, we would arrive at an empirical formula $H_{15750}N_{310}O_{6500}$ $C_{2250} Ca_{63}P_{48}K_{15}S_{15}Na_{10}Cl_6Mg_3Fe_1$. The relative numbers of atoms in this differ from the composition by weight because atoms have different masses.

The composition of an alien life-form would depend on two key factors: First is the element that forms the "skeleton" of its macromolecules. All life discovered so far is based on carbon, which can form long chains to which other elements bind. The most likely alternative building block for macromolecules would be silicon, phosphorus, or nitrogen. Second is the solvent for the biochemical reactions that drive the body. The most likely alternative to water is probably ammonia (NH_3) because it can dissolve most organic molecules. It is also liquid well below water's freezing point and is prevalent in space. So an alien life-form might be based on silica and ammonia.

LAURI SUORANTA

The chemical elements in an adult human are distributed in various molecular and atomic species. An accurate formula could be expressed in the standard form: $7 \times 10^{25}H_2O + 9 \times 10^{24}C_6H_{12}O_6 + 2 \times 10^{24}CH_3(CH_2)_{14} + \ldots$ and so on. However, such a series would fill a book and we cannot possibly identify all species. Metabolism, defined as the chemical and energy exchanges in a living body, means that any such chemical formula is continually changing.

Having a chemical formula for a process can be useful. If we find all the elements and determine all the mathematical expressions applying to them, the whole process can be determined. But this is not the whole story. Life is characterized by extensive adaptive self-regulation of its own structural order, and utilizes feedback control. An organism uses its resources in its own emergent way. The chemical reactions work, but how they are brought together is a matter of emergent control systems. This means that not only is it impossible to write an accurate formula for a human being; it is unnecessary and can be misleading to try. Life is what it does with chemical species, not just which ones it is made from.

I guess the same would go for any alien life-form we might encounter. We spend considerable time searching the electromagnetic spectrum to detect signals, and we receive a lot of signals. But how will we know if any of them are life? Only, I suppose, if they show the characteristic of life: I'm in control, and I'm not solely a bottom-up, deterministic chemical process.

JOHN WALTER HAWORTH

⚙ Zzzzapppp

When people die of electric shock, what kills them—current or voltage?

Kyle Skotzke

It is the current through the heart region that causes most deaths from electric shock. The effect depends on duration of exposure and also varies between individuals. The frequency of home electrical power—around 50 or 60 hertz—is very dangerous, and currents of only a few tens of milliamps at such a frequency can cause the heart to fibrillate. It pulses at a much higher rate than normal and fails to pump blood to the brain; death follows in a few minutes.

Because the body has electrical resistance, the current flowing in it depends on the voltage. The current also depends on the dampness of the skin and where on the body the current enters and leaves. It is therefore very difficult to come up with a safe voltage for all circumstances. This is being attempted at the moment by the International Electrotechnical Committee (IEC) working group on electric shock, but the number of variables makes simple recommendations difficult.

There are other mechanisms that can cause death from electric shock. One of these is muscular contraction. If a current passes through the chest it can inhibit breathing and lead to asphyxia. A current in the head can affect the respiration center in the brain, again leading to asphyxia. Once more, it is current, rather than voltage, that is the critical factor.

Most people who receive an electric shock survive. This is not because they are particularly strong but because there is usually some factor that reduces the current, such as resistance from clothing or shoes, or the length of the shock. A ground-fault circuit interrupter, often touted as a safety device, is useful to shorten the duration of a shock but does not prevent the shock from occurring in the first place.

In short, it is a function of current and time that kills.

N. C. Friswell
International Electrotechnical Committee working
group on electric shock
Horsham, West Sussex, UK

Damage from an electric shock varies with current. However, except in the case of superconductors, voltage is needed to drive this current, so the distinction is a little artificial. If the resistance of the human body were constant, then voltage would be an equally valid yardstick. But the resistance varies according to a number of factors.

For example, dry skin offers an electrical resistance of 500,000 ohms. Yet wet skin reduces this to 1,000 ohms—only double the resistance of salty water. So being soaked to the skin leaves us more vulnerable to harm.

The path of the current is critical. This is why standing on insulating material and doing electrical work with one hand behind your back, so that a grounded current will pass not across your chest but down to your feet, reduces the chance that a current will pass through your heart. The heart can stop if current passes through it, and we can suffer severe burns as electrical energy is converted to heat.

Alternating current is said to be four or five times as dangerous as direct current, because it induces more severe muscular contractions. It also stimulates sweating, which lowers the skin's resistance, increasing the current passing through the person. Sixty cycles a second happens to be the most harmful range.

Thomas Edison tried to take advantage of this fact when, in 1886, New York state established a committee to replace hanging with a more humane form of execution. He employed Harold Brown to invent the electric chair, powered by the alternating current that was favored by his rivals in the race to commercialize electricity distribution. If it were used to kill criminals, Edison hoped that potential customers would shun alternating current in favor of the direct-current system he had developed. Sadly for Edison, this interesting piece of marketing turned out to be unsuccessful because AC proved to be cheaper and can be stepped up to higher voltages to be transported more efficiently over great distances.

MIKE FOLLOWS

Electricity kills by delivering energy where it is not wanted. Energy is the product of voltage, current, and time. It could be lethal when delivered as low as 100 microamps at a few volts if sent directly to the heart, or about 30 milliamps at a few hundred volts from one hand to another. In both cases the problem arises if the shock disorganizes the electrical activity of the heart to make the ventricles fibrillate. Of course, the solution to this problem is to deliver another shock, from a defibrillator, if you have one handy.

Electrical energy can kill you in other ways. The electric chair appears to kill by asphyxiation, because it causes uncontrolled contraction of the muscles of respiration. It also cooks its victims a bit, but does not seem to reliably produce either ventricular fibrillation or rapid loss of consciousness from current through the brain. In other circumstances, large currents that pass through the body without causing instant death can cause horrible, deep burns. These can, of course, kill you more slowly. Finally, a high-voltage discharge can set fire to your clothes or blow you off the electricity pylon you might be working on; either effect can be fatal.

MIKE BROWN .

◙ Take Your Pick

Is it coincidence a human finger fits exactly into a human nostril? If not, why does my mother tell me not to do it?

JACK WALTON (AGE 9)

Your mother may not approve, but there is a way to clear your nose without sticking anything inside it. It's called the "snot rocket." Just push against the side of one nostril to close it off, take a deep breath, close your mouth, and exhale as hard and sharply as you can through your other nostril. You'll be amazed how fast the contents shoot out. Just make sure you tilt your head away from your body to avoid peppering yourself.

Nose-clearing tactics like the snot rocket mean there is no life-or-death reason for the coevolution of digging digits and large, inviting nostrils. After all, nose blockage is easily managed by breathing through your mouth. In fact, a blocked nose is really a problem only if something gets lodged near your nasal bones, where it is dangerously close to your brain. That is a region where human fingers are too pudgy to be of any use. A rather thrilling story of a primatologist, some tweezers, and an engorged Ugandan tick comes to mind.

Sexual selection might have favored the relationship of finger to nostril if, say, females in the Pleistocene preferred mating with males who picked their noses, or if males and females picked each other's noses in a courtship ritual. However, that would be taking reciprocal grooming a little far.

So we must conclude that, yes, it is mere coincidence that your fingers fit so nicely into your nostrils. I doubt the made-for-each-other argument is going to change your mom's opinion of rhinotillexomania. I suggest you demonstrate the snot rocket instead and see what she says.

Holly Dunsworth

Organs commonly correspond in size and shape to other organs with which they must function. Conspicuous examples include the male and female sexual organs of many insects and some mammals, the mouths of baby marsupials and their mothers' nipples, and—in many animals—elongated claws or toes that have been adapted for grooming. However, a mismatch need not mean that the organs cannot work together. For example, the mammalian female birth channel can obviously accommodate the passage of young that are far bigger than the male sexual organ. Apertures often expand or shrink to fit the organs that they match.

Conversely, it does not follow that because an organ fits an aperture, the fit is other than coincidental. There are some other places your finger would fit into that your mom would tell you firmly to leave alone, especially if you were in public.

You have five sizes of finger and two nostrils, so to get some sort of fit does not demand much of a coincidence. Nor is there any obvious reason why there should be any selective pressure to adapt nostrils to finger-reaming. More likely, nature intended us to dribble snot just as elephant seals do. The fine art of nose-picking is just another adventitious setback for intelligent design.

JON RICHFIELD

While I agree that an expertly executed unilateral snot rocket is indeed a thing of beauty and wonder, I would caution against Holly Dunsworth's suggestion that you "exhale as hard and sharply as you can" from one nostril. My developing expertise in this technique as a schoolboy was brought to an abrupt halt by a burst sinus and severe nosebleed.

DUNCAN HANNANT
PROFESSOR OF LARGE ANIMAL IMMUNOLOGY
UNIVERSITY OF NOTTINGHAM
LOUGHBOROUGH, LEICESTERSHIRE, UK

I'd also like to add a word of warning that this method isn't particularly hygienic, and could spread any number of diseases. Snot rockets should really be practiced only when you are by yourself.

BRON

2   Feeling OK?

◈ Catch Your Death

Is there any connection between being cold and catching a cold? If not, why is there so much folklore about catching a cold if you sleep uncovered or in a draft?

ANTONIS PAPANESTIS

No, there is no connection. The erroneous association developed for several reasons.

The viruses that cause colds spread faster in the winter because people spend more time inside, where they are closer together.

Because people close the windows in winter, air contaminated by virus particles is not diluted by "fresh" air from the outdoors. This makes it easier for the virus to spread.

The cold, dry air of winter makes the mucous membranes in the nose swell. This produces the "runny nose" we often incorrectly associate with an infection caused by a cold virus.

The experience of catching a chill and getting a cold is actually the reverse of the correct order of things. The chill is often the first sign of fever that is the result, not the cause, of the infection by the cold virus.

MARK FELDMAN

Studies have shown that there is no correlation between environmental temperature and suffering from colds. The origin of the old wives' tale that predicts colds, flu, or pneumonia after exposure to cold temperatures is the short period of fever that precedes the distinctive symptoms of these illnesses. These periods of fever make the patient feel cold and shivery. Shortly after developing other symptoms, the patient then associates the illness with having "caught cold." Indeed, the flu is called influenza from the belief that it was caused by the "influence" of the elements. The fact that isolated researchers living in Antarctica never catch colds confirms that these are caught from people and not from "cold."

PEDRO GONZALEZ-FERNANDEZ

There is actually less chance of your catching a cold in the cold. The virus known as the common cold dies in cold and needs warmth (say, the cosy indoors of a home beside the fire started to keep out the cold) to thrive.

ESPERANDI

It's Not

Sorry, but I had to ask: why is nasal mucus often green?

DAVID TANNER

Of all the body cavities in contact with the outside world, the nose is probably one of the most hospitable: it is warm, very well aerated, and moist; and it supplies unlimited quantities of bacterial food secreted continuously by the nasal mucosa (mucus contains quantities of glycoprotein and dissolved salts). In other words it is an ideal breeding ground for bacteria, which are always present.

Many of the common bacteria associated with humans are colored: *Staphylococcus aureus* is a golden yellow, for example, and *Pseudomonas pyocyanea* (to give it its older but more explicit name) is a shade of blue. Normally these and the multitude of other organisms that are inhaled continuously into the nose are flushed out by runny mucus, which is swallowed. The bacteria are usually digested.

However, if a situation arises where the flow of mucus slows down and then becomes much thicker in response to an infection of any kind, then the bacteria, in their ideal home, can multiply and produce the colored mucus described. This, as many parents know, is one of the less endearing characteristics of babies and young children!

And, by the way, if you're still wondering where the green color comes from, remember what happens when you add blue to yellow.

Laurie North

Your previous correspondent suggested that the green color is caused by a combination of golden-yellow *Staphylococcus aureus* and blue *Pseudomonas pyocyanea*. This is a rather tenuous belief. While the eighth edition of Bergey's *Manual of Determinative Bacteriology* (Williams & Wilkins, Baltimore, 1974, page 222) still held *P. pyocyanea* "commonly isolated from wound, burn and urinary tract infections" to be the causative agent of "blue pus," the cause of the green color of pus or nasal mucus is more general.

Green pus (or green nasal mucus) is caused by iron-containing myelo-peroxidases and other oxidases and peroxidases used by polymorphonuclear (PMN) granulocytes (neutrophils). These short-lived phagocytizing leukocytes avidly ingest all sorts of bacteria and inactivate them by oxidative processes, involving the iron-containing enzymes above. The resulting breakdown product (comprising dead PMNs, digested bacteria, and used enzymes), pus, contains significant amounts of iron, which gives it its greenish color.

C. J. van Oss
Department of Microbiology
State University of New York, Buffalo
and
J. O. Naim
Department of Surgery
Rochester General Hospital, New York

Nasal mucus isn't always green. Nasal mucus produced at the beginning of a cold is clear and is produced in response to tissue damage caused by the invading rhinovirus. It turns green a few days into the infection as neutrophils respond to clear away the cellular debris and secondary bacterial infection sets in.

JULI WARDER

Polymorphonuclear leukocytes are equipped with a number of enzymes, the most potent of which is peroxidase. This same peroxidase is also found in horseradish, giving it a distinctive green color and a sharp (if fleeting) bite, as anyone who has tried Japanese wasabi paste can confirm. English horseradish sauce loses its color, owing to oxidation of this labile enzyme on exposure to air. However, authentic wasabi is always served fresh.

Sorry if this response puts some readers off their sushi.

STEVE FLECKNOE-BROWN

❖ Ouch!

What causes the pain induced when a piece of tinfoil touches a tooth filling?

SIMON ODDY

The questioner is advertently replicating a famous experiment first performed by Luigi Galvani in 1762.

When two dissimilar metals are separated by a conducting liquid, a current will flow between them, and this current can be used to stimulate nerves.

This is exactly what happens when silver foil appears to touch the amalgam of a filling. A thin film of saliva actually separates foil from filling, and because saliva is a reasonable electrolyte, containing various salts, a current will flow between the tooth and the filling. As the filling is close to the dental nerve, the current will stimulate it, causing pain.

Galvani carried out his experiment with frogs' legs and metal probes, but the effect was the same—they twitched!

CHRIS QUINN

⬡ Speaker's Throat

What are the bodily changes that cause us to have a dry throat when we are nervous?

HOWARD FOSS

You get a dry mouth during public speaking because when you are nervous the body is set into the "fight or flight" state. This is caused by an activation of the autonomic nervous system. It is seen throughout the animal kingdom, and has evolved to help the animal deal with dangerous situations—when escaping from predators, for example.

The nerves are selectively activated, depending on how important they are for the response. Because eating is not considered to be important at this time—you want to get the hell out of the place—the nerves to your mouth that control the salivary glands are suppressed, so your mouth dries up. In addition, your pupils dilate and the blood vessels to your muscles and heart are enlarged in order to get the blood to the most important organs needed for whatever drastic action is necessary.

M. SCOTTEN

This is tied to the "fight or flight" reaction. In a tense or dangerous situation, your body shuts down all unnecessary functions, including the digestive system. Your saliva glands are a part of it. You don't need to digest your last meal if a lion is trying to make you his next. This is also where butterflies in the stomach come from.

BILL ISAACSON

Light Night

When you have a heavy cold, your nose runs all day but stops as soon as you fall asleep at night. What mechanism turns off the flow of mucus? Could a drug be developed to do the same and therefore alleviate the worst effect of the cold? And could such a drug also help to reduce the spread of colds?

PETER ROONEY

This phenomenon is merely a matter of gravity. When you settle down to go to sleep, the action of lying on your back means that the mucus being secreted in your nasal passages, rather than running down and out of your nose, instead travels toward the back of the throat, and you swallow it unconsciously.

If you sleep on your side, just one nostril (the lower one) becomes blocked, and if you want to unblock your nose, it is usually sufficient merely to alter your orientation—so if you are lying down, stand up; and if you are standing up, lie down. This reverses the direction of the mucus flow and clears the blockage.

ALEXANDRA MCKENZIE JOHNSTON

The runny nose that occurs while we're awake is caused by head position. When we sleep, almost all the postnasal drip drains down the throat because we are lying down, sometimes on our side, and sometimes on our back. I suffer from this problem, and, after testing my theory using a massage table that had a face-hole opening, I started using pillows to position my head so my nose points downhill.

Test it yourself using pillows, leaving a gap so that you can rest your head facedown while leaving a space for your mouth and nose. Oh, and have some handkerchiefs ready: you'll probably find that your nose runs copiously all night.

HANK ROBERTS

The questioner is mistaken in his belief that the runny discharge is contagious. Most colds are contagious from two to four days after exposure to the virus—that is, usually before the symptoms appear. Once you start to suffer from a runny nose, the infection is well under control and the amount of virus present in the nasal secretion has diminished.

The most common way to catch a cold is through hand contact with the virus rather than from people with runny noses. Colds are more easily transmitted via a hard surface that other people have touched, such as a doorknob or computer mouse.

Instead of avoiding people with runny noses, it is far more important to make sure that you do not touch your eyes, nose, or mouth with your hands after they have been in contact with various objects in either your home or your office.

DAVID GIBSON

✧ All in the Mind

Discussion of the placebo effect in the testing of therapies always seems to focus on positive placebo effects. Are there negative placebo effects?

PETER GRANT

Placebos are substances, such as sugar or dummy pills, with no pharmacological properties. They are widely used as a control in experiments to test the effect of medicines, and are made to look and smell the same as the drug being tested. Subjects are not told whether they are receiving the actual medicine or the placebo.

How the placebo effect works is still controversial, but it is widely believed that the effect is psychological rather than physiological: the benefit occurs because people believe that the pill they are taking should cause positive effects. The effect has also been attributed to conditioning: patients expecting the effects of a drug will then experience them.

Take the example of placebos used in tests on analgesic drugs. An explanation for the placebo mechanism in this case is that it involves the release of opiate-like pain-relieving chemicals in the brain. One study found that pain was reduced by a placebo medication that the patients believed was a pain reliever, but that the effect ceased when the patients were given a drug that counteracts the effect of opiates.

The negative effects of placebos are called *nocebo* effects, nocebo being Latin for "I will harm." Patients receiving dummy pills sometimes experience side effects such as anxiety and depression. These are thought to be associated with the person's expectations of adverse effects of the treatment as well as conditioning. It was reported in one trial that women who believed they were prone to heart disease were nearly four times as likely to die from heart disease as women with similar risk factors who had no such belief.

Placebos pose an ethical dilemma. They work primarily because a doctor deceives patients into believing that they are receiving an active medicine, while in fact depriving them of any such medicine. If they also suffer nasty side effects through the nocebo effect, this arguably makes things worse.

IAN SMITH

Yes, there are negative placebo effects, or nocebo effects. Nocebos, like placebos, cause a physical effect, though not primarily through a physical mechanism. It seems likely that their effect comes from the patient's belief. It's "think sick, be sick" for the nocebo, as opposed to "think well, be well" for the placebo.

The kind of patient most likely to experience the nocebo effect of a given drug has a history of vague, difficult-to-diagnose complaints and is sure that whatever therapy is prescribed will do little to battle the problem. Those low expectations are invariably met. The nocebo effect also affects the outcome of operations. Surgeons are wary of people who are convinced that they will die. Studies have been carried out on people undergoing surgery who say that they wish to die to be reunited with a loved one. Almost all of these people do die.

There is very little research on nocebos, mostly for the ethical reason that physicians ought not to induce illness in patients who are not sick. And changing ethical standards have made it difficult to repeat some of the classic nocebo experiments. The most recent medical review article on nocebo effects was published in 2002 by Arthur Barsky and others (*Journal of the American Medical Association*, volume 287, page 622).

ROSS FIRESTONE

There are negative placebo effects. Their best-known manifestations are voodoo and other "magical thinking" involving curses. Such practices almost invariably include some mechanisms for letting the victim know that he or she has been cursed, and this is the main, possibly the only, requirement for success.

STEVEN REITCI

⚙ Expect Pain in Knee Area

I damaged my knee ligaments in a skiing accident about two years ago. Ever since then I have had what I describe as a "weather-forecasting knee." Before it rains I always experience pain in my knee. This happens in both summer and winter and does not seem to be related to humidity. I am not the only person to have reported this. Why does my knee hurt before it rains and, more interestingly, how does it know? How does it detect the onset of rain?

DEBBIE REID

Plenty of studies have looked at pain associated with weather, especially in people suffering from arthritis. These show that there is a real effect, but, oddly, there has been little research into what causes the pain—Ed.

The human body can be viewed schematically as a clump of gelatin-filled balloons mounted on a stick. Undamaged tissue—fat, muscle, or bone—is relatively elastic and will expand and contract when subjected to changes in atmospheric pressure. Scar tissue, in contrast, is very stiff and dense, and does not expand or contract appreciably within the range of normal atmospheric fluctuation.

Imagine that several of the balloons in your hypothetical body clump were glued together and then the surrounding pressure was lowered. The balloons would expand, so the glued-together area—representing the scar tissue—would distort and pull as a result. In living tissue this effect results in nerve stimulation and a rapid onset of pain. These persist until the pressure normalizes or the scar eventually stretches to relieve the distortion. The process may take hours or days.

I occasionally mystify my office staff by announcing in the morning: "It's going to be a busy day for drop-in patients." They never know how I can predict the 20 or 30 who will call with a severe increase in pain from surgery or an old injury. I'd rather my staff think I have magical abilities than confess I read the weather report.

Immersion in a hot tub and gentle exercise may ease the pain. Waiting for the weather to change works, too, and here in east Texas it usually changes before you can fill the tub.

STEVEN BALLINGER

It may sound like an old wives' tale that wet weather aggravates arthritis, but in the 1960s a rheumatologist named Joseph Hollander built an experimental climate chamber to test the claim. He found that high humidity combined with low barometric pressure—the meteorological situation before it rains—is indeed associated with joint pain or stiffness.

One explanation is that the change in weather makes injured ligaments swell, and the nerves around the joint sense this as pain. Another is that air within the joint may expand when barometric pressure drops, again causing the nerves to report pain.

A recent experiment by Japanese scientists demonstrated that back pain associated with changes in barometric pressure is linked to the vacuum phenomenon, in which gases build up in the spaces between the vertebrae (*Journal of Spinal Disorders and Techniques,* volume 15, page 290). Such bubbles form as the disks between the vertebrae deteriorate, and are more common in older people. They can also form in other joints.

Avoid pain by keeping your knee dry and warm. And, of course, you are now more qualified than most to work as a local TV weather forecaster.

FRANK WONG

One explanation for the weather-predicting knee is "bone bruising"—bleeding and edema caused by microscopic fractures of trabecular, or porous, bone. Some studies have found that these are relatively common after ligament injury to the knee.

Maybe changes in atmospheric pressure could change the volume of the edema in the bone and produce pain. If so, two predictions might be made: an MRI scan will show bone bruising, and the patient's ability to predict weather should decrease as the injury heals.

PETER HALLAS

3 Plants and Animals

◈ Birdzzz

Why do birds never fall off their perches when sleeping? Do they, in fact, sleep?

GRAEME FORBES

Birds have a nifty tendon arrangement in their legs. The flexor tendon from the muscle in the thigh reaches down over the knee and continues down the leg, around the ankle, and then under the toes. This arrangement means that, at rest, the bird's body weight causes the bird to bend its knee and pull the tendon tight, closing the claws.

Apparently this mechanism is so effective that dead birds have been found grasping their perches long after they have died.

ANNE BRUCE

Yes, birds do sleep. Not only that, but some do it standing on one leg. And, even more surprising, some may be hypnotized into sleep. My mynah bird was.

If you wish to try it, bring your eyes close to the cage, and use the hypnotist's "your eyes are getting heavier" principles (not spoken) on your own eyes. Act as if you are gradually falling asleep and the bird will follow you, finally holding one leg up under its belly, tucking its head under its wing, and falling into a deep sleep.

What's more, most owners of pet birds know that all you need to do to make your pet fall asleep is to cover the cage with a blanket to simulate night.

DAVID LECKIE

Birds do sleep, usually in a series of short "power naps." Swifts are famous for sleeping on the wing. Since most birds rely on vision, bedtime is usually at night, except for nocturnal species, of course.

The sleeping habits of waders, however, are ruled by the tides rather than the sun.

Some other species are easily fooled by artificial light. Brightly lit city areas can give songbirds insomnia. A floodlighted racetrack near my home gives an all-night dawn effect on the horizon, causing robins and blackbirds to sing continuously from 2 A.M. onward. Unfortunately, I don't know whether it tires them out as much as it does me . . .

ANDREW SCALES

◈ Tank Madness

Following a recent bereavement we would like to know why fish jump out of small aquariums.

ROWAN WHITE AND VICKY
UNIVERSITY OF EAST ANGLIA, NORWICH, UK

Fish jumping out of small tanks are quite a common problem for enthusiasts, and this is the reason why some owners choose to have a glass cover on the top of their aquarium.

There are several theories as to why fish might jump from a small aquarium. It has been suggested that in the wild they use this method to attempt to rid themselves of ectoparasites.

Although the questioners did not mention the gender and species mix of the fish in their aquarium, it is possible that their fish could have been leaping to avoid predators or unpleasant interactions with other creatures, or even to show off to their conspecific fish, in some previously unknown courtship or territorial ritual.

In the meantime, my sincere condolences for your loss.

R. ROSENBERG

To captive fish, the air on the other side of the aquarium glass looks like water. And in fish lore, the water is always cleaner on the other side.

JOHN CHAPMAN

⚙ Baaa-rmy

Why do sheep always run in a straight line in front of a car and not to the side?

ALED WYNNE-JONES

Sheep and other animals run ahead of cars because they do not realize that cars cannot climb grassy banks. Ancestral sheep were pursued by wolves and big cats. If an animal tries to turn aside some yards from the hunter, the pursuing animal sees what is happening, makes an easy change of course, and intercepts the victim, which is presenting its vulnerable flank.

If, however, the prey dodges at the last minute, the outcome is different. The hare is the master of this strategy: as the greyhound reaches out with its jaws, the hare makes a quick evasive turn to one side and the dog overshoots or, with luck, tumbles head over heels.

The instinctive response of a sheep or a hare to an approaching car is at least not as maladaptive as that of the hedgehog.

CHRISTINE WARMAN

Herbivores are killed by predators who normally grab them by the throat while running alongside, so the prey will always do its best to keep a potential threat behind its tail, swerving as the predator attempts to overtake. That's why a kangaroo, seeing a car drawing alongside, will jump onto the road right ahead in order to keep the car directly behind it, and often get run over in the process. As long as a car proceeds in a straight line behind a sheep, the sheep will try to outrun it in a straight line.

G. Carsaniga

Sheep are much underrated. They don't merely run in a straight line—they run straight for a while, then dive to the side. This is not woolly thinking; it's perfectly logical. Sheep loose in the road are usually confined to country areas, where roads are bounded by steep verges, cliffs, hedges, fences, and ditches. The sheep recognizes that if it cannot beat the car on the flat, it stands no chance whatsoever up a bank, so it attempts to outrun the vehicle down the road.

What happens then is that the vehicle slows, and when the vehicle reaches a speed that is slow enough for the sheep to think it might beat the car over the obstructions at the side of the road, it swerves. And since most of the time this action is proved correct (most vehicles don't follow sheep off the road), the sheep carries on behaving in this way. QED, by sheep logic.

Clearly, this is a much more successful approach to road safety than that shown by humans, who rarely try to out-pace the oncoming car. They tend to dive to the side of the road. Since more people are run over than sheep, one can conclude we have much to learn from sheep logic.

William Pope

Sheep, being clever animals with an instinctive grasp of psychology, know that most drivers, though enjoying an occasional kill as long as they can use the "it jumped in front of me, there was nothing I could do" excuse, are not so depraved as to run something down deliberately. Thus running in a straight line has a distinct advantage over veering to the side.

Erik Decker

NATIONAL INSTITUTE OF ANIMAL HUSBANDRY, DEPARTMENT OF
CATTLE AND SHEEP, TJELE, DENMARK

⊙ Fried Fish

*My young neighbor asked me what happens when lightning
strikes water. Do all the fish die, and what happens to the
occupants of metal-hulled boats?*

CHRIS COOPER
KEMPSTON, BEDFORDSHIRE, UK

When a bolt of electricity, such as a lightning bolt, hits a watery surface, the electricity can run to earth in a myriad of
directions.

Because of this, electricity is conducted away over a
hemispheroid shape, which rapidly diffuses any frying power
possessed by the original bolt. Obviously, if a fish was directly hit by lightning, or close to the impact spot, it could
be killed or injured.

However, a bolt has a temperature of several thousand
degrees and could easily vaporize the water surrounding the
impact point. This would create a subsurface shock wave
that could rearrange the anatomy of a fish or deafen human
divers over a far wider range—tens of yards.

If people in a metal-hulled boat were close enough to feel
the first effect they would be severely buffeted by the second.
Besides, metal hulls conduct electricity far better than water,
so a lightning bolt would travel through the ship in preference to the water.

ANDREW HEALY

When lightning strikes, the best place to be is inside a conductor, such as a metal-hulled boat, or under the sea (assuming you are a fish).

In the nineteenth century, the physicist Michael Faraday
showed that there is no electric field within a conductor. He

demonstrated this by climbing into a mesh cage and then striking artificial lightning all over it. Everybody except Faraday was surprised when he climbed out of the cage unhurt.

ERIC GILLIES
UNIVERSITY OF GLASGOW, UK

✷ Blowfish

Fish don't fart. Why is this?

CHRISTINE KALIWOSKI

The writer probably thinks that fish don't fart because she has not seen a string of bubbles issuing from a fish's vent.

However, fish do develop gas in the gut, and this is expelled through the vent, much as in most animals. The difference is the packaging.

Fish package their excreta into a thin gelatinous tube before disposal. This includes any gas that has formed or been carried through digestion. The net result is a fecal tube that either sinks or floats, but as many fish practice coprophagia, these tubes tend not to hang around for too long.

DEREK SMITH

I have on several occasions witnessed my cichlids passing wind, to the displeasure of my plub eel.

This seems to be a result of their taking in too much air while wolfing down flaked foods floating on the surface of the water. If the air was not expelled it would seriously affect their balance.

PETER HENSON

Most sharks rely on the high-density lipid squalene to provide them with buoyancy, but the sand tiger shark, *Eugomphodus taurus,* has mastered the technique of farting as an extra buoyancy device. The shark swims to the surface and gulps air, which it swallows into its stomach. It can then fart out the required amount of air to maintain its position at a certain depth.

ALEXANDRA OSMAN

Cold Feet

Why do Antarctic penguins' feet not freeze in winter when they are in constant contact with the ice and snow? Years ago I heard on the radio that scientists had discovered that penguins had collateral circulation in their feet that prevented them from freezing, but I have seen no further information on or explanation of this. I have asked scientists studying penguins about this, but none could give an answer.

SUSAN PATE

Penguins, like other birds that live in a cold climate, have adaptations to avoid losing too much heat and to preserve a central body temperature of about 104°F. The feet pose particular problems, since they cannot be covered with insulation in the forms of feathers or blubber, yet have a big surface area (similar considerations apply to cold-climate mammals such as polar bears).

Two mechanisms are at work. First, the penguin can control the rate of blood flow to the feet by varying the diameter of arterial vessels supplying the blood. In cold conditions the flow is reduced; when it is warm, the flow increases. Humans can do this too, and that is why our hands and feet become white when we are cold and pink when warm. Control is very sophisticated and involves the hypothalamus and various nervous and hormonal systems.

However, penguins also have "countercurrent heat exchangers" at the top of the legs. Arteries supplying warm blood to the feet break up into many small vessels that are closely allied to similar numbers of venous vessels bringing cold blood back from the feet. Heat flows from the warm blood to the cold blood, so little of it is carried down to the feet.

In the winter, penguins' feet are held a degree or two above freezing—to minimize heat loss while avoiding frostbite. Ducks and geese have similar arrangements in their feet, but if they are held indoors for weeks in warm conditions, and then released onto snow and ice, their feet may freeze to the ground, because their physiology has adapted to the warmth and this causes the blood flow to the feet to be virtually cut off and their foot temperature falls below freezing.

JOHN DAVENPORT
UNIVERSITY MARINE BIOLOGICAL STATION
MILLPORT, ISLE OF CUMBRAE, UK

I cannot comment on the presence or absence of collateral circulation, but part of the answer to the question about penguins' cold feet has an intriguing biochemical explanation.

The binding of oxygen to hemoglobin is normally a strongly exothermic reaction: an amount of heat (DH) is released when a hemoglobin molecule attaches itself to oxygen. Usually the same amount of heat is absorbed in the reverse reaction, when the oxygen is released by the hemoglobin. However, as oxygenation and deoxygenation occur in different parts of the organism, changes in the molecular environment (acidity, for example) can result in overall heat loss or gain in this process.

The actual value of DH varies from species to species. In Antarctic penguins things are arranged so that in the cold peripheral tissues, including the feet, DH is much smaller than in humans. This has two beneficial effects. First, less heat is absorbed by the birds' hemoglobin when it is deoxygenated, so the feet have less chance of freezing.

The second advantage is a consequence of the laws of thermodynamics. In any reversible reaction, including the absorption and release of oxygen by hemoglobin, a low temperature encourages the reaction in the exothermic direction, and discourages it in the opposite direction. So at low temperatures, oxygen is absorbed more strongly by most species' hemoglobin, and released less easily. Having a relatively modest DH means that in cold tissue the oxygen affinity of hemoglobin does not become so high that the oxygen cannot dissociate from it.

This variation in DH between species has other intriguing consequences. In some Antarctic fish, heat is actually released when oxygen is removed. This is taken to an extreme in the tuna, which releases so much heat when oxygen separates from hemoglobin that it can keep its body temperature up to 30°F above that of its environment. Not so cold-blooded after all!

The reverse happens in animals that need to reduce heat due to an overactive metabolism. The migratory water hen has a much larger DH of hemoglobin oxygenation than the humble pigeon. Thus the water hen can fly for longer distances without overheating.

Finally, fetuses need to lose heat somehow, and their only connection with the outside world is the mother's blood supply. A decreased DH of oxygenation by the fetal hemoglobin when compared with maternal hemoglobin results in more heat being absorbed when oxygen leaves the mother's blood than is released when oxygen binds to fetal hemoglobin. Thus heat is transferred into the maternal blood supply and is carried away from the fetus.

CHRIS COOPER AND MIKE WILSON
UNIVERSITY OF ESSEX, COLCHESTER, UK

Flying Fins

Why do flying fish fly? Is it to escape predators, or to catch flying insects, or as a more efficient means of getting around than swimming? Or is there some other entirely different reason?

JULYAN CARTWRIGHT

The usual explanation for flight in flying fish is as a way to escape predation, particularly from fast-swimming dolphin-fish. They do not fly to catch insects; flying fish are largely oceanic and flying insects are rare over the open sea.

It has been suggested that their flights (which are actually glides, because flying fish do not flap their "wings") are energy-saving, but this is very unlikely, as the vigorous take-offs are produced by white, anaerobic muscle beating the tail at a rate of 50 to 70 beats per second, and this must be very expensive in terms of energy use.

Flying fish have corneas with flat facets, so they can see in both air and water. There is some evidence to suggest that they can choose landing sites. This might allow them to fly from food-poor to food-rich areas, but convincing evidence of this is lacking.

There seems to be little doubt that escape from predators is the major purpose of their flight, and this is why so many fly away from ships and boats, which they perceive to be threatening.

JOHN DAVENPORT
UNIVERSITY MARINE BIOLOGICAL STATION
MILLPORT, STRATHCLYDE, UK

Strictly speaking, the flying fish does not fly; it indulges in a form of powered gliding, using its tail fins to propel it clear of the water. It sustains its leap with high-speed flapping of its oversized pectoral fins for distances of up to 325 feet. The sole purpose of this activity seems to be to escape predators. If one can manage to tear one's eyes away from the magic of the unexpected and iridescent appearance of a flying fish, a somewhat more substantial fish can often be seen following its flight path just below the surface.

TIM HART

I have seen whole schools of flying fish become airborne as they try to escape tuna which are hunting them, and minutes later have seen the school of tuna attempt similar acrobatics as dolphins move in for their supper of tuna steaks.

A morning stroll around the decks of an oceangoing yacht will often provide a frying pan full of flying fish for breakfast. Presumably they are instinctively trying to leap over a predator (in this case the boat), but as they don't seem to be able to see too well at night they land on the deck. They rarely land on board during the day. Most alarmingly, they will land in the cockpit, and even hit the stargazing helmsman on the side of the head.

DON SMITH

◈ Not Mush Room

Near where I live there are toadstools growing through the pavement, the surface of which they have displaced in fairly large chunks. What mechanism allows toadstools—essentially very soft and squashy items—to push through 2 inches of asphalt?

JOHN FRANKLIN

The toadstools forcing their way up through asphalt are probably ink cap mushrooms *(Coprinus)* growing on buried plant debris. They are pushing upward because their stalks function as vertical hydraulic rams.

The upward pressure comes from the turgor pressure of the individual cells making up the wall of the hollow stalk of the mushroom. Each individual cell grows as a vertical column by inserting new cell wall material uniformly along its length.

The major structural component of the cells is a shallow helical arrangement of fibers of chitin winding around the axis of the cell. These chitin fibers are embedded in matrix materials, making the wall material like a carbon fiber composite. Chitin is an exceptionally strong biopolymer (also used by insects for their exoskeletons) and gives immense lateral strength to the fungal cell wall, so that internal pressure is confined as a vertical column. Water enters the cell by osmosis, and the resulting turgor pressure provides the vertical force that allows the mushroom to push up through the asphalt.

This phenomenon was first investigated 75 years ago by Reginald Buller, who measured the lifting power by loading weights onto a mushroom that was elongating inside a glass tube. He calculated an upward pressure of about two-thirds of an atmosphere.

The cells have a gravity-sensing mechanism that keep the mushroom exactly vertical. A mushroom that is put on its side will rapidly reorient to grow vertically again.

GRAHAM GOODAY
UNIVERSITY OF ABERDEEN, UK

Two inches of asphalt is nothing to the muscular mushroom. One large shaggy ink cap *(Coprinus comatus)* discovered in Basingstoke lifted a 30- by 24-inch paving stone 1½ inches above the level of the pavement in about 48 hours.

Historically, mushrooms often sprang up in foundries, supposedly from horse manure used in preparing loam for casting, and were often reported as having lifted heavy iron castings. Presumably these would have been some type of field mushroom such as *Agaricus campestris*. Whatever the species, the mechanism by which the force was exerted is likely to be the same: hydraulic pressure.

As Buller found, the exquisite and fragile *Coprinus sterquilinus* exerts an upward pressure of nearly half a pound with a stem one-fifth of an inch thick, so it is not surprising that more robust species can tear the tarmac.

RICHARD SCRASE
MUSHROOM MAKERS
BATH, SOMERSET, UK

⊙ Them!

I have been amazed to see ants emerge seemingly unharmed after being zapped in the microwave, usually after hitching a ride on my coffee cup. They seem to run around quite happily while the microwave is in operation. How can they survive this onslaught?

JUDITH KELLY

The answer is quite simple. In a conventional microwave, the waves are spaced a certain distance apart, because that is all that is needed to cook the food properly. The ants are so minute that they can dodge the rays and so survive the ordeal.

LI YAN

The phenomenon that the ants take advantage of is that microwaves form standing waves within the oven cavity.

So in some places in the oven space, the energy density is very high, whereas in others it is very low. This is why most ovens have turntables to ensure that cooking food is heated evenly throughout.

This standing wave pattern can be observed by putting a static tray of marshmallows in the microwave, and heating for a while. The result will be a pattern of cooked and uncooked marshmallows. The standing wave pattern, however, varies according to the properties and position of any material within the oven, such as a cup of water.

The ant will experience this pattern as hot or cold regions within the oven and can thus locate the low-energy volumes. If the ant is in a high-field region, its high surface-area-to-volume ratio allows it to cool down more quickly than a larger object while it searches for a cold spot.

It is a common myth that microwaves are too big to heat small objects. The fallacy of this reasoning has been demonstrated by chemists such as myself who employ microwave heating in their work. Certain types of catalyst consist of microwave-absorbing particles—typically of submicron size—dispersed throughout an inert support material. There is convincing evidence that microwaves are capable of heating only the tiny catalyst particles.

A. G. Whitaker

There is very little microwave energy near the metallic floor or walls of the oven. The electromagnetic fields of microwaves are "shorted" by the conducting metal, just as the amplitudes of waves in a skipping rope, swung by a child at one end but tied to a post at the other, are reduced to nothing at the post. An ant crawling on the rope could ride out the motion near the post, but might be thrown off nearer the middle.

For a quick demonstration of this, place two pats of butter in a microwave in two polystyrene coffee cup bottoms, one resting on the floor, the other raised on an inverted glass tumbler. Be sure to place a cup of water in the microwave as well. When the oven is turned on, the raised butter will melt long before the butter on the floor.

Charles Sawyer

How About Gnat?

How is it possible for gnats to fly in heavy rain without being knocked out of the air by raindrops?

L. Pell

A falling drop of rain creates a tiny pressure wave ahead (below the raindrop). This wave pushes the gnat sideways and the drop misses it. Flyswatters are made from mesh or have holes on their surface to reduce this pressure wave; otherwise, flies would escape most swats.

Alan Lee

The world of the gnat is not like our own. Because of the difference in scale, we can regard a collision between a raindrop and a gnat as similar to that between a car moving at the same speed as the raindrop (speed does not scale) and a person having only one-thousandth the usual density—for example, that of a thin rubber balloon of the same size and shape. A balloon is easily bounced out of the way, and would burst only if it was crushed up against a wall.

TOM NASH

Beeline

My girlfriend tells me it is impossible to explain how the bumblebee flies. Apparently it defies the laws of physics. Is this true?

TORBJØRN SOLBAKKEN

The infamous case of the flightless bumblebee is a classic example of carelessness with approximations. It stems from trying to apply a basic equation from aeronautics to the flight of the bee. The equation relates the thrust required for an object to fly to its mass and the surface area of its wings. In the case of the bee, this gives an extremely high value—a rate of work impossible for such a small animal. So the equation apparently "proves" bees cannot fly.

However, the equation assumes stationary rather than flapping wings, making its use in this case misleading. Of course if equations fail in physics there is always empirical observation—if a bee looks as if it is flying, it most probably is.

SIMON SCARLE

Conquering Conkers

I was once advised by a friend that the way to strengthen a conker before a conker fight is to bake it. From my childhood I remember being told that the way to improve your conkers was to pickle them in vinegar. Which method produces winning conkers and why?

BY E-MAIL, NO NAME SUPPLIED

The simplest and best way to harden conkers is to put them away in a drawer until the following year. However, if they were not attached to a string when new and soft they will have to be drilled.

Both my children and my grandchildren played with my leftover conkers, some of which were 50 years old. They have never been defeated.

F. GRISLEY

I have been in conker fights for about 50 years and I always soak them in vinegar. This hardens them into champion conkers. I was content with this method until a few years ago when I was beaten by someone who had smeared his conker in Oil of Olay. Apparently, this made the conker more malleable, allowing it to absorb the impact of my prize pickled nut.

MICHAEL DUTTON

Neither baking nor pickling in vinegar is an effective way to strengthen conkers. Baking makes chestnuts brittle, which means they can be knocked off their string with a single blow. Pickling rots the inside. Varnishing, another technique used, is also ineffective (and readily detectable).

In fact, no intervention is necessary to toughen a conker. Simply avoid using chestnuts from the current season (conkers is obviously an autumn pursuit) and use old ones instead. The older they are, the harder they are. Such conkers are readily identified—instead of having a glossy chestnut brown skin they will look dull and dark, perhaps even black. And finally, make the hole for threading the string as narrow as possible.

NICK AITCHISON

The best way to make a conker invincible is either to leave it for a year and use it in the season after you found it or, to speed up the process, bake it. Put all your conkers in the oven at 250°F for about two hours.

Do not leave them any longer; otherwise, the flesh inside the conker will become charred and weak. Even if the heat breaks the shell of your conker, the flesh will be rock-hard.

Do not put your conkers in vinegar. Although vinegar hardens the shells, the flesh will soften up if there is any gap in the shell, making the conker useless.

PATRICK WIGG

Every conkerer disagrees with every other on how best to produce an invincible nut. As such disputes are an essential part of the sport, we leave the question with the totally contradictory answers given above—Ed.

I was intrigued by all the dialogue about conkers, and I gather that they are chestnuts, somehow attached on a string. What is it you British do with the chestnuts? I suspect it doesn't involve consumption.

JENNIFER HOLTZMAN

What kind of game is conkers? Is it like a pillow fight with rock-hard chestnuts? Sometimes we yokels in the former colonies need a little edification . . .

JAY KANGEL

Conker fighting appears to be a mainly British pursuit so we realize we have to enlighten our international readership. Conkers are the hard fruit of the horse chestnut tree. These are collected in autumn, removed from their spiky casing, and left to mature. A hole is then drilled in the conker, and a string threaded through. The full rules of genuine competition are complex but the game as played by schoolchildren (and overgrown schoolchildren) is between two opponents each with one conker. One player dangles a conker by the string, holding it steady, while the opponent swings his or her conker on its string and attempts to strike the hanging conker. Players take turns doing this until one conker is so damaged that it is dislodged from its string. The winner is obviously the player with the intact conker. Naturally, the stronger and harder the conker, the more chance of success.

This is perhaps further proof, if it were needed, of the British obsession for devising eccentric and meaningless methods of competition—Ed.

The Australian game is called "bullies" and is played in a manner similar to the way the English play conkers.

I know of its being played at least between 1900 and 1970 in western New South Wales and South Australia and before the advent of the TRS-80 and Commodore 64.

As most fair-dinkum Aussies with a bush background will know, the bully is the seed of the quandong tree. Also known as the wild peach, this widely distributed bush tree requires a host tree to survive and fruits annually, producing a tart, bright red fruit, up to 2 inches in diameter.

This fruit has been an Australian delicacy in pies and jams for many years but is only now becoming more commercially available. The stone from the fruit is perfectly round and dimpled like a golf ball. It is usually about 1 inch in diameter, with the requisite internal nut and as hard as a stone. Drilling was a difficult task, with good bullies causing the demise of many a parent's drill bit.

Local rules governed the length of the string and the size of the playing circle. Ownership of losing bullies was not an issue, as my recollection is that losers shattered.

No heat treatment was needed or desired—fire and heat are necessary for the germination of most Australian bush seeds, so heating would certainly weaken the bully. I am unaware of an international challenge in this enthralling sport, but I did consider proposing it for the Sydney Olympics in 2000. My money would have been on the bullies from the colonies.

JIM BILLS

My brother was disqualified from the school conker championships for having a conker that was vacuum-impregnated with epoxy resin.

J. McINTYRE

Eggsactly

Why are most eggs egg-shaped?

MAX WIRTH

Eggs are egg-shaped for several reasons. First, this shape enables them to fit more snugly together in the nest, with smaller air spaces between them. This reduces heat loss and allows best use of the nest space. Second, if the egg rolls, it will roll in a circular path around the pointed end. This means that on a flat (or flattish surface) there should be no danger of the egg's rolling off, or out of the nest. Third, an egg shape is more comfortable than a sphere or a cylinder for the bird while it is laying (assuming that the rounded end emerges first).

Finally, the most important reason is that hens' eggs are the ideal shape for fitting into egg cups and the egg holders on the fridge door. No other shape would do.

ALISON WOODHOUSE

Most eggs are egg-shaped (ovoid) because an egg with corners or edges would be structurally weaker, besides being distinctly uncomfortable to lay. The strongest shape would be a sphere, but spherical eggs will roll away and this would be unfortunate, especially for birds that nest on cliffs. Most eggs will roll in a curved path, coming to rest with the sharper end pointing uphill. There is in fact a noticeable tendency for the eggs of cliff-nesting birds to deviate more from the spherical, and thus roll in a tighter arc.

JOHN EWAN

Eggs are egg-shaped as a consequence of the egg-laying process in birds. The egg is passed along the oviduct by peristalsis—the muscles of the oviduct, which are arranged as a series of rings, alternately relax in front of the egg and contract behind it.

At the start of its passage down the oviduct, the egg is soft-shelled and spherical. The forces of contraction on the rear part of the egg, with the rings of muscle becoming progressively smaller, deform that end from a hemisphere into a cone shape, whereas the relaxing muscles maintain the near hemispherical shape of the front part. As the shell calcifies, the shape becomes fixed, in contrast to the soft-shelled eggs of reptiles, which can resume their spherical shape after emerging.

Advantages in terms of packing in the nest and in the limitation of rolling might play a role in selecting individuals which lay more extremely ovoid eggs (assuming the tendency is inherited) but the shape is an inevitable consequence of the egg-laying process rather than evolutionary selection pressure.

A. MacDiarmid-Gordon

✺ Lucky Mark

Local birds tend to eat little black insects. So how come they void themselves on me from a great height with a white and annoyingly conspicuous product?

M. ROGERS

It is a common misconception that the white droppings produced by birds are feces.

In fact, they are urine. Birds excrete uric acid rather than urea because it is an insoluble solid. This way they avoid wasting water when urinating—just one of their adaptations for a good power-to-weight ratio.

GUY COX
UNIVERSITY OF SYDNEY, AUSTRALIA

The white material that makes up the droppings of birds, and indeed many reptiles, is their urine.

The more primitive vertebrates excrete toxic nitrogenous waste relatively directly, having masses of water at their disposal with which they can dilute substances such as ammonia.

However, birds and reptiles—at least lizards and snakes, with whose droppings I'm very familiar—are different. It would appear that the conversion of their toxic nitrogenous waste products into a relatively insoluble one that can then be formed into a paste was an evolutionary adaptation. This enabled them to lead a terrestrial rather than aquatic life, and even to live in ecological niches where water is scarce.

In such niches it is particularly important not to have to find extra water with which to dilute toxic waste products and flush them from the system, so birds and lizards solved this by evolving to produce a paste of insoluble and relatively nontoxic uric acid.

Interestingly, birds that consume a lot of roughage in their diets, such as the heather-eating grouse and ptarmigan, produce droppings that are very similar to guinea-pig feces. Only here and there among the droppings is it possible to make out the telltale white patches of their urine, so copious is their production of feces.

PHILIP GODDARD

Your previous correspondents omit one fact, oviparity. The evolution of insoluble excreta has nothing to do with a

"good power-to-weight ratio" or the ability to "live in ecological niches where water is scarce."

It evolved because all birds and many reptiles begin their
life inside an egg. Even heavy egg-laying amniotes that live in
water as adults, such as penguins and crocodiles, must survive this early phase without poisoning their shelled enclo-
sure with any water-soluble metabolites.

ÖRNÓLFUR THORLACIUS

They do so from a great height because from a lower height
it's just too easy to hit the target—no challenge at all. The
deposit needs to be white so that, from said great height,
they can see where it lands and who it hits.

S. B. TAYLOR

Red or White?

Why is red meat red and white meat white? What is the dif-
ference between the various animals that makes their flesh
differently colored?

TOM WHITELEY

Red meat is red because the muscle fibers that make up the
bulk of the meat contain a high content of myoglobin and
mitochondria, which are colored red. Myoglobin, a protein
similar to hemoglobin in red blood cells, acts as a store for
oxygen within the muscle fibers.

Mitochondria are organelles within cells that use oxygen
to manufacture the compound ATP, which supplies the energy
for muscle contraction. The muscle fibers of white meat, by
contrast, have a low content of myoglobin and mitochondria.

The difference in color between the flesh of various animals
is determined by the relative proportions of these two basic
muscle fiber types. The fibers in red muscle fatigue slowly,
whereas the fibers in white muscle fatigue rapidly. An active,

fast-swimming fish such as a tuna has a higher proportion of fatigue-resistant red muscle in its flesh, whereas a much less active fish such as the plaice has mostly white muscle.

TREVOR LEA

The color of meat is governed by the concentration of myoglobin in the muscle tissue, which produces the brown coloring during cooking.

Chickens and turkeys are always assumed to have white meat, but free-range meat from these species (especially that from the legs) is brown. This is because birds reared in the open will exercise and become fitter than poultry grown in restrictive cages. The fitter the bird, the greater the ease of muscular respiration, and hence the increased myoglobin levels in the muscle tissue, making the meat browner.

All beef is brown because cattle are allowed to turn around in fields all day, but pork is whiter because pigs are lazy.

T. FILTNESS

 # Job Swap

If polar bears were transferred to Antarctica, could they survive? And would penguins survive in the Arctic?

RICHARD DAVIES

Polar bears would probably survive in the Antarctic, and the Southern Ocean around it, but they could devastate the native wildlife. In the Arctic polar bears feed mainly on seals, especially young pups born on ice floes or beaches. Many of the differences in breeding habits between Arctic and Antarctic seals can be interpreted as adaptations to evading predation by bears.

Polar bears would find plenty of fish-eating mammals and birds around Antarctica. Penguins would be particularly vulnerable because they are flightless and breed on open ground, with larger species taking months to raise a single chick. Bears can run only in short bursts, but they could catch a fat, sassy penguin chick or grab an egg from an incubating parent.

In the Arctic polar bears hunt mainly on the edge of the sea ice, where it is thick enough to support their weight but thin enough for seals to make breathing holes. The numerous islands off the north coast of Canada, Alaska, and northwest Europe provide plenty of suitable habitats. The Antarctic continent is colder, with only a few offshore islands, so bears would probably thrive at lower latitudes in the Southern Ocean than in the Arctic.

We can only hope that nobody ever tries what the questioner suggests. Artificially introduced predators often devastate indigenous wildlife, as it is not accustomed to dealing with them. This occurred with stoats in New Zealand, foxes and cats in Australia, and rats on many isolated islands.

Large, heavy animals would also trample the slow-growing, mechanically weak plants and lichens of the Antarctic. For instance, Norwegian reindeer have decimated many native plants in South Georgia, an island in the South Atlantic Ocean, since they were introduced 80 years ago.

C. M. POND
DEPARTMENT OF BIOLOGICAL SCIENCES,
OPEN UNIVERSITY,
MILTON KEYNES, BUCKINGHAMSHIRE, UK

While, as far as I know, no one has ever been stupid enough to introduce polar bears into the Antarctic, there have been at least two practical attempts to transplant penguins to the Arctic.

The original "penguin" was in fact the late great auk *(Pinguinus impennis)*, once found in vast numbers around northern shores of the Atlantic. Although not a relative of southern hemisphere penguins, it was very similar in appearance, and filled much the same ecological niche as penguins, particularly the king penguins of the subantarctic region.

With any attempt to introduce an alien species, there must actually exist an appropriate ecological niche for it to fill, and that niche must be vacant. For the most part, the ecological niches occupied by penguins in the south are filled by the auk family to the north. But the demise of the great auk in the mid-nineteenth century at the hands of hungry whalers created not only a vacancy that one of the larger penguins might neatly slot into, but also a potential economic demand for the penguin's fatty meat and protein-rich eggs.

It was perhaps the possible economic opportunities that prompted two separate bids to introduce penguins into Norwegian waters in the late 1930s. The first, by Carl Schoyen of the Norwegian Nature Protection Society, released groups of nine king penguins at Røst, Lofoten, Gjesvaer, and Finnmark in October 1936. Two years later, the National Federation for the Protection of Nature, in an equally bizarre operation, released several macaroni and jackass penguins in the same areas, even though these smaller birds would clearly find themselves competing directly with auks or other native seabirds.

The outcome was unhappy for the experimenters and, most particularly, for the penguins. Among those whose fate is known, one king was quickly dispatched by a local woman who thought it was some kind of demon, while a macaroni died on a fishing line in 1944, although from its condition it had apparently thrived during its six years in alien waters.

And it soon became obvious that the real reason why any attempt to fill the ecological gap left by the great auk was destined to fail was the very reason that the niche was vacant in the first place—such large seabirds could not happily coexist with a large and predatory human population. Of course, it is the steadily increasing human presence in the far south that is now threatening penguins in their native habitat.

HADRIAN JEFFS

◉ How Does He Smell?

Why are dogs' noses black?

RACHEL COLIN (AGE 11)

While a majority of dogs have black noses, not all do. The noses of dogs such as vizslas and Weimaraners match their coat colors—red and silver, respectively—and it is not unusual for puppies of any breed to start out with pink noses that then darken as the animal matures. I had a Shetland sheepdog that retained pink on the insides of her nostrils for the whole of her life.

Dogs have most likely developed black noses as a protection against sunburn. While the rest of the dog's body is protected by fur, light-colored noses are exposed to the full force of the sun's rays. Pink-nosed dogs, hairless breeds, and dogs with very thin hair on their ears need to be protected with sunscreen when they go out of doors, just as humans sometimes do, or they risk the same sorts of cancers and burns.

In addition, dog breeders have long singled out a black nose as the only acceptable color for many breeds. Though this is based on nothing more than an aesthetic preference, it still serves as a selective influence for people breeding pedigreed dogs. This adds a bit of human-directed evolution to what was already a natural tendency toward black noses.

JULIA ECKLAR

Black nose leather contains the skin pigment melanin, specifically in its dark brown or black eumelanin forms. Melanocytes, the cells that produce the raw material, secrete it into the skin cells, and the sun then darkens it further. Melanin in skin cells protects the DNA in cells from mutations caused by ultraviolet radiation from the sun.

JON RICHFIELD

4 Food and Drink

⚙ Banana Armor

The skin of bananas in the fridge turns brown faster than those in a room, but the fruit is still edible. I thought the browning was oxidation, but if so why does it happen faster in the cold?

ALUN WALTERS

I wouldn't recommend putting bananas in the fridge to keep them fresh. Like all living organisms bananas adjust the composition of their cell membranes to give the right degree of membrane fluidity for the temperature at which they normally live. They do this by varying the amount of unsaturated fatty acids in the membrane lipids: the colder the banana, the greater the level of unsaturated fatty acid and the more fluid the membrane at a given temperature. If you chill the fruit too much, areas of the membrane simply become too viscous and the cell membranes lose their ability to keep the different cellular compartments separate. Enzymes and substrates that are normally kept apart therefore mix.

Overripe fruit kept out of the fridge goes brown by the same mechanism, but in this case membrane breakdown occurs as a part of the general senescence of the tissue. In chilled commercial storage injury is, in fact, a big problem with tropical fruits, whereas temperate fruits like apples and pears can happily be stored at temperatures near freezing. I wonder, therefore, whether bananas stored in the fridge really taste as good as those left out. Incidentally, since tomatoes are a semitropical fruit I wouldn't suggest you keep them in the fridge either.

ALISTAIR MACDOUGALL
INSTITUTE OF FOOD RESEARCH
NORWICH, UK

Although many fruits are stabilized by refrigeration, most tropical and subtropical fruits (bananas in particular) exhibit chill injury. The ideal temperature for bananas is 56°F. Below 48°F spoilage is accelerated by the release of enzymes and the skin can blacken overnight as the banana fruit and skin soften. The enzymes leak from cellular storage sites and the leakage is caused by increased membrane permeability. This is mediated by ethylene gas, which controls ripening and response to chill injury, and by such events as attack by parasites.

The two enzymes that break down the main polymers responsible for plant structure are cellulase and pectinesterase. These break down cellulose and pectin respectively. The breakdown of starch by amylase-type enzymes is also involved in softening banana fruit tissue.

The blackening of the skin is caused by the release of another enzyme, polyphenyl oxidase (PPO). This is an oxygen-dependent enzyme which polymerizes naturally occurring phenols in the banana skin into polyphenols similar in structure to melanin formed in suntanned human skin.

PPO is also inhibited by acid, and this is why lemon juice is used to prevent browning in apples. Bananas are low in acidity, and this may be one reason why they darken so quickly. Finally, blackening of the skin can be slowed by coating the banana with wax to exclude oxygen.

M. V. WAREING

Further to your previous answers: yes, the browning is an oxidation reaction. Yes, it is initiated by cooling. But no, the low temperature itself does not speed up the oxidation reaction in bananas.

Bananas like hot climates, and their cell membranes are damaged in the fridge. Membrane damage lets the phenolic amines such as dopamine, which are normally present inside the vacuoles of banana skin cells, leak out and encounter oxidizing enzymes (polyphenol oxidases) elsewhere in the cell. The dopamine can then be oxidized by atmospheric oxygen to form brown polymers, which may act as a defensive barrier. Once started by the cold-induced membrane damage, the browning reaction is promoted by warming.

For an extreme demonstration, put a banana skin in the freezer for a few hours. It stays creamy white because, although the membranes are shattered by freezing, oxidases cannot work at such low temperatures. Now let it thaw overnight at room temperature: the skin will go pitch-black as the dopamine is oxidized. A control banana skin kept at room temperature overnight stays whitish because the vacuolar membranes remain intact.

STEPHEN FRY
UNIVERSITY OF EDINBURGH, UK

⊙ White-Water Drinking

Why do anisette-based drinks, such as Pernod and Sambucca, turn white when water is added to them?

ALEXANDER HELLEMANS

Anisette-based drinks rely on aromatic compounds called terpenes for their flavor. These terpenes are soluble in alcohol, but not in water. The 40 percent or so alcohol in the drink is enough to keep the terpenes dissolved, but when the drink is mixed with water they are forced out of solution to give a milky-looking suspension.

Absinthe, a similar drink, which is based on wormwood and is now banned in some countries because of its toxicity, gives a more impressive green suspension. Terpenes are responsible for a lot of the harsher plant scents and flavors, including lemongrass and thyme.

THOMAS LUMLEY

⬡ Transparent Rock

How do you get transparent ice? Ice from my freezer always contains bubbles. I've used filtered and boiled water but the ice is never like that seen in ads for Scotch.

PHILIP SUSMAN
MONASH UNIVERSITY, VICTORIA, AUSTRALIA

Ice made in domestic freezers is inevitably cloudy because of the dissolved air that tap water contains (around 0.003 percent by weight). As the water in the ice trays drops below the freezing point, crystals form around the edges of the compartments. These are pure ice and they contain very little air because the solubility of air in ice is very low and the liquid left behind can still hold it in solution.

Once the concentration of air in the liquid reaches 0.0038 percent by weight and the temperature has dropped to 31.9957°F, the liquid can contain no more air and a new reaction begins. As the water freezes, the air is forced out of solution. The natural state of air at the temperature and pressure involved here is a gas, so it forms bubbles in the ice.

Commercial ice machines produce attractive, clear ice by passing a constant stream of water past freezing metal fingers, or over freezing metal trays. This freezes out a fraction of the water while the rest of it is discarded before the concentration of air gets too high. When the ice is thick enough, the metal fingers or trays are warmed to release their crystal-clear ice that is good enough to film.

Alas, without an ice machine, the questioner may have to make do with cloudy ice cubes.

ANDREW SMITH

Water has its highest density at around 7°F. Below that the water gets less dense as it approaches freezing point.

Air bubbles form in the ice when the cooling of the water is too rapid, causing one part of the water to be at a temperature different from other parts. Ice is usually formed at the top of the water first because the warmer and denser water sinks beneath the ice layer that begins to form there.

Additionally, the top layer is usually the part that is in contact with the cold environment. This is similar to what happens in a frozen lake. The various expansion rates of different parts of the water will invariably create air bubbles that cannot escape because of the ice sheet above.

To avoid ice bubbles, the trick is to cool the water very slowly so that there is no large temperature gradient which can cause widely different expansion. Cooling it slowly also allows air to have sufficient time to move through the liquid and evaporate before it is trapped by the solid ice.

HAN YING LOKE

Your failure to achieve clear ice even after boiling the water may depend on what precautions you took in addition to filtration and boiling. For instance, gas can still be dissolved if water is poured after boiling has finished, and hard water may also need deionizing to rid it of gas.

It also helps to exclude air while the water is cooling, perhaps with plastic wrap. Try to achieve zone freezing by cooling the water slowly from the top down, maybe in a polystyrene container with only plastic wrap over the top. Do not use a thermos bottle, because the glass is far too fragile.

Although the amount of gas that freezes out is unaffected by the method of freezing, zone freezing starts by giving a nice, thick chunk of clear ice. As freezing progresses, murky ice appears, and then you can stop the process.

JON RICHFIELD

Water contains dissolved gases. When it freezes, the gas is forced out of it to form bubbles that are then trapped in the ice, making it look opaque.

In order to get transparent ice you should use warm rather than cold water, since warm water contains less dissolved gas. Also, try to reduce the power of your freezer in order to allow time for the gases to diffuse out just before freezing. I have tried this and it works very well.

GABRIEL SOUZA

I'm afraid that your correspondent has been seduced by a professional photographer who uses hand-carved Perspex "ice" cubes in ads for Scotch because they don't melt under the studio lights. If he looks very carefully, he might also see the tiny glass bubbles on the meniscus of other drinks—these won't disappear at the wrong moment either.

MARTIN HASWELL

Over the Top

Sparkling wine or beer poured into a dry glass froths up. If the glass is wet this does not happen. Pour some sparkling wine into a glass so that it froths up to the rim, let the bubbles subside, and you can then pour the rest of the wine quickly, knowing it will not froth over the top. Why?

H. SYDNEY CURTIS

Beer, sparkling wine, and other fizzy drinks are liquids which are supersaturated with gas. Although thermodynamics favors the gas bubbling out of the dissolved state, bubble formation is unlikely, since bubbles must start small.

The pressure of these tiny bubbles can reach about 30 atmospheres in a bubble only a fraction of an inch in diameter. Because the solubility of gases increases with increasing pressure (Henry's law), the gas is forced back into solution as quickly as it comes out.

Bubbles can form around dust particles, surface irregularities, and scratches. These nucleation sites are hydrophobic and allow gas pockets to grow without first forming tiny bubbles. Once the gas pocket reaches a critical size, it bulges out and rounds up into a properly convex bubble whose radius of curvature is sufficiently large to prevent self-collapse . . .

D. P. MAITLAND
DEPARTMENT OF PURE AND APPLIED BIOLOGY
UNIVERSITY OF LEEDS, WEST YORKSHIRE, UK

In addition, there is a cascade effect. If the quantity of bubbles reaches a certain critical number per unit volume, this in itself constitutes a physical disturbance and results in the release of yet more bubbles.

Nucleation may be precipitated by a variety of imperfections. Minute crystals of salts (such as calcium sulfate) may remain if the glass has been left to dry by evaporation after being washed in hard water. Tiny cotton fibers may be left behind if the glass has been dried with a dish towel. Dust particles may have settled on the glass if it has been left standing upright for any length of time. And tiny scratches will be present on the inside surface of all but brand-new glasses.

Once the inside of the glass is wet, any salt crystals will have dissolved and any cotton fibers will no longer function as centers of nucleation. Most of the dust particles and all of the scratches will, of course, still be there. However, these will have been coated with liquid, and the fresh carbonated liquid can reach them only very slowly, by diffusion. Bubbles will still be produced, but at a rate that is too slow for the cascade effect to come into play. As a result, the drink will not froth over.

ALLAN DEEDS

To demonstrate the above, take a glass and thoroughly coat the inside with an oil, which is a more efficient surface covering agent than water. Then add a less expensive carbonated drink such as seltzer. The effervescence will be nil or minimal. Add a few million centers of nucleation from a large spoonful of granulated sugar and the effervescence will be volcanic.

RONALD BLENKINSOP

Thanks to modern production techniques, today's glasses are of such good quality that some manufacturers build in deliberate imperfections, especially in beer glasses, in order to generate enough bubbles to maintain the head on the top of your tipple.

TONY FLURY

⊙ Slice Crisis

What is the irritant that causes my eyes to water when I am slicing onions? Is there any way to prevent this?

STEPHEN MITCHELL

Onions and garlic both contain derivatives of amino acids that in turn contain sulfur. When an onion is sliced, one of these compounds, S-1-propenylcysteine-sulphoxide, is decomposed by an enzyme to form the volatile propanthial S-oxide, which is the irritant or lacrimator.

Upon contact with water—in this case your eyes—the irritant hydrolyzes to propanol, sulfuric acid, and hydrogen sulfide. Tearfully, the eyes try to dilute the acid. However, it is these same sulfur compounds that form the nice aroma when onions are being cooked.

To prevent watering eyes, I would suggest one of the following: stop using onions (but you would lose the tasty aroma); wear goggles (you would look slightly silly); slice the onion under water (you will wash some of the aroma out); before slicing the onion, wash it, and keep it wet.

BERND EGGEN

To attempt to reduce the severity of watering eyes you must allow the maximum possible time for the irritant to disperse before it comes into contact with the eyes. The most obvious way of achieving this is to stand as far away from the onion as your arms will allow. It also helps if you are not standing over the onion, but back from it.

Another way of reducing the amount of irritant reaching your eyes is to breathe through your mouth. This means that you do not create a current of air which flows up to your nose and onward to the eyes, carrying the irritant with it; instead, the air is either directed into the lungs when you are breathing in or forced away from the face when you are breathing out.

In order to ensure that you breathe through your mouth, hold a metal spoon lightly between your teeth. There will be space for the air to enter and escape, and while our mouths are open we breathe preferentially through them, rather than our noses. I find that holding the spoon upside down works best, although I don't know what scientific reason there could be for this.

C. BURKE

I have found that wearing contact lenses prevents eye irritation when I am chopping onions.

ELAINE DUFFIN

A slice of lemon should be placed under the top lip while one is slicing. One does not look particularly attractive, but it does prevent the eyes from watering.

SHEILA RUSSELL

I suggest the old tip of holding a sugar cube between your teeth to absorb the irritant. It does work, as do sulfur matches, though very few people use these now.

MICHEL THURIAUX

Hold a piece of bread—say a quarter of a slice—between the lips as you slice onions. This was taught to my family in Tanzania in the early 1960s by our cook, Victor Mapunda, from Malawi.

JOHN NURWICK

Question of Class

We are told to let red wine breathe before drinking it to improve the flavor. At the risk of sounding like a Philistine, I wonder: wouldn't it be quicker to pour it into a cocktail shaker, shake for 10 seconds, and let the bubbles subside?

CHRIS JACK

Wine is left to breathe to allow the volatile and aroma-bearing substances to start evaporating, so that we may enjoy the bouquet. Shaking a drink is completely different. An agitated drink incorporates gas, letting oxygen reach as much liquid as possible. This oxidized liquid provides a very different taste.

For some drinks this taste may be pleasant. However, if you oxidize wine you obtain vinegar, which, I suspect, is not the flavor you wish to taste. Therefore, there is a genuine reason for drinks being "shaken, not stirred" or vice versa, depending on what you have in your glass.

PAUL MAVROS
ARISTOTLE UNIVERSITY
THESSALONIKA, GREECE

The reasons usually given for decanting red wines have changed during the past few years. This is because of two developments: one in wine-making technology and the other in wine tastes.

The original reason for decanting was to separate the wine from organic particulates formed by precipitation, and aggregation from tartaric acid, tannin compounds, original microparticulates present in the pressed grape juice and proteinaceous material that is formed during maturation of the wine.

Because these particulates are small to minuscule in size and of a density not much higher than the wine itself, Stokes's law predicts that they will sink back to the bottom only extremely slowly should they be suspended by careless motion of the bottle.

This is the reason for those magnificent mechanical decanting machines, which allow precisely controlled tilting of the bottle to reduce suspension of particulates.

A very different reason for decanting lies in aerating the wine to hasten the release of the secondary elements of its nose. While traditional old wines may actually lose some of their olfactory elements through intense aeration and become stale quickly, decanting for aeration parallels the development of taste in younger wines or wines elevated in oak casks, with associated different weighting of primary and secondary smells.

In Italy, where many progressive vintners have been experimenting with new assemblages and methods of elevation, decanting often means pouring the contents of a bottle straight down into a decanter, generating lots of chaotic turbulence with an intense mixing of air and wine.

In the hands of a self-confident wine waiter this process can look flamboyantly spectacular. As a logical development of this reason for decanting, some modern Italian glass decanters have a flattened shape that allows for the maximum air-wine interface, giving further aeration.

OLIVER STRAUB

It is generally recognized that red wine should be drunk at ambient temperature and, since it is often stored in a relatively cold room or location (near the floor), the most important aspect of the so-called breathing process is to raise the wine's temperature.

However, the ambient temperature in the UK is often a little low, and red wine is usually best if drunk at about 86°F. Placing a bottle of red wine in a microwave oven for 50–60 seconds (depending on the season) on high power will produce the required effect without your having to resort to allowing the wine to breathe before consumption, but do not forget to remove the foil capsule and the cork. The alternative concept of shaking the wine in a cocktail shaker will result in the formation of various oxidation products including vinegar, which will have a negative effect on the flavor.

M. V. WAREING

Only chemists drink red wine at a temperature of 86°F. Our wine experts suggest a temperature of around 63°F—Ed.

◪ One or Two?

Expert advice says that you should use freshly drawn water every time you make a pot of tea or coffee. Why is this? What is wrong with water that has been boiled twice? Can anyone tell the difference?

IVOR WILLIAMS

The reason that freshly boiled water is more effective for making tea than water boiled twice is that the fresh water has a higher oxygen content. This should result in a tastier brew, because more tea will be extracted from the tea leaves.

This can be easily demonstrated by placing a measured amount of tea leaves in two glass tumblers and adding freshly boiled water to one and repeatedly boiled water to the other. Examination of both tumblers after 3 minutes will reveal a much stronger brew from the freshly boiled water.

J. R. Stafford
Marks & Spencer
London, UK

I was told as a child that the reason for using freshly drawn water to make tea was that the dissolved oxygen made the tea taste better. Water which has been standing or, worse, had been boiled contains less dissolved oxygen. The British Standard 6008, which describes in great detail how to make a cup of tea, says that the water must be freshly boiling but does not say anything about its being freshly drawn. It also says that the milk should be put in the cup first to avoid scalding it.

As this British Standard is identical to International Standard ISO 3103, the supplementary question is: why can't I get a decent cup of tea abroad?

N. C. Friswell

The traditional explanation for making tea with freshly boiled water is that prolonged boiling drives off the dissolved oxygen, making the tea taste "flat." My own experiments with water simmered for an hour versus freshly boiled water produced little perceptible difference, even though high-quality leaf tea was used and brewed for 5 minutes.

I would be surprised if the difference was of the slightest practical importance for tea made by dunking a tea bag, especially if the water had been boiled merely twice.

David Edge

I see that at least one reader remains unconvinced about the need to use freshly boiled water for tea.

Once, during an emergency overseas, we were instructed to boil all drinking water for several minutes. It didn't seem to affect the tea. However, we decided that it would be a good idea to use a domestic pressure cooker to raise the water temperature to beyond boiling point to sterilize the water thoroughly. This was fine when the water was used for drinking or cooking, but when we tried using it for making the tea the result was absolutely dreadful.

On the other hand, I have drunk tea at an altitude of 6,800 feet, where, of course, the boiling point is lower than 212°F, but I noticed no difference in the taste. Nor did my tea-planter hosts make any comment on the point.

Pressure-cooked water apart, I think the length of time the tea is allowed to infuse is a more critical factor.

A. C. ROTHNEY

A. C. Rothney may be surprised to hear that his/her shiny pressure cooker probably caused his/her nasty tea. Dissolved aluminum in the water, not the higher temperature to which the water had been subjected, is the reason the tea tasted awful. In the days when most kettles were made of aluminum they carried instructions to prepare the new kettle by repeatedly boiling fresh water and then discarding it. Only then should the first pot of tea be made with fresh water. During these repeated boilings, a patina of dull oxide built up inside the kettle and prevented the water from dissolving the pure aluminum.

LORNA ENGLISH

The preference for fresh water in making tea has little to do with oxygen but is related to dissolved metal salts (mainly calcium and magnesium bicarbonates, sulfates, and chlorides) which are present as impurities in tap water and which affect the color and taste of tea.

The effect of metal salts on the color of tea can be demonstrated by comparing a brew made with freshly boiled pure water (deionized or melted freezer frost) with tea made with freshly boiled tap water. The salts in tap water give a darker brew, which is cloudier as a result of precipitated insoluble salts such as tannates.

Boiling tap water destabilizes the bicarbonates (so-called

temporary hardness), which precipitate out as insoluble carbonates on cooling (this is why a kettle furs up with time). In hard-water areas, where more dissolved salts are present, repeated boiling and cooling will remove sufficient calcium and magnesium salts, although boiling for a long time without cooling has less effect.

There are three reasons why repeatedly boiled and cooled water can produce less palatable tea. First, some of the precipitated carbonate remains in suspension, even after reboiling, as a white scum (particularly noticeable in new plastic kettles), and this taste is more marked than bicarbonates dissolved in water—especially when the scum interacts with the tea.

Second, the salts in the water that are not destabilized by boiling (so-called permanent water hardness) are gradually concentrated by evaporation, producing unpleasant flavors.

Finally, traces of metals, such as iron and copper, can accumulate in repeatedly boiled water, and these can interact with oxygen and reducing agents in the tea (phenols) by complex redox reactions to produce further effects on flavor.

M. V. WAREING

As a caffeine addict, I suffer severe headaches if I go more than a day without my cups of tea. To conserve fuel on hikes lasting a number of days, I tried leaving a tea bag in a bottle of cold water for a few hours. It worked. Not only did it give me my fix of caffeine, but it tasted like tea, albeit cold tea. I haven't yet tried making such a cold infusion, then heating it in a microwave, but it should prove quite drinkable.

SYD CURTIS

The truth runs counter to A. C. Rothney's ideas.

My father was a tea taster and faultless at detecting whether we had boiled the water for too long. How did he do it?

Hard waters (and most waters do have some mineral salts in solution, causing hardening) brew more slowly than soft or alkaline waters. If you boil hard water for considerably longer than the standard half a minute or so, more of the dissolved salts deposit themselves on the inside of the kettle. The emerging water is then softer than expected and softer than the tea taster balanced the tea blend for. It will brew faster and with a darker color than usual.

Tea manufacturers ensure constant performance by balancing their blend differently for sale in different water areas of the country, even if the brand label is the same. Hard water can be artificially softened with a pinch of bicarbonate of soda, but the dramatic darkening of the color and change of flavor are unacceptable to most people—including tea tasters.

BERNARD HOWLETT

✥ Twister

Here in Zimbabwe we buy milk in plastic packages. Most people cut a tiny piece off the edge of the package to pour the milk. I have noticed that when the milk leaves the package under pressure it exits in a corkscrew or a spiral fashion. Of course, other liquids would behave in a similar fashion. What forces operate to allow an unchanneled liquid to follow this path? I have noticed that the smaller the opening in the package the greater the amount of twisting in the path followed by the milk.

DAVID WHITE

The corkscrew effect you notice is just the bottom end of the whirlpool that is occurring inside the carton as the milk exits. The force that causes it is usually called the Coriolis force. This is responsible for all whirlpooly stuff you might find. Milk cartons and bottles give you the same effect, but it is less noticeable because of the shape of the cross section of their openings.

When the milk leaves the contents under pressure because you are squeezing the carton while pouring, you effectively increase the liquid's speed. This increases the Coriolis force— which is proportional to the speed of an object in a rotating inertial frame, as well as the frame's angular velocity and the distance from the object to the axis of rotation. This gives a tighter corkscrew. In effect, milk under pressure screws up.

JOHN LENTON

The twisting of the stream of milk coming out of the package has more to do with the shape of the hole (usually a long, thin one), the difference in pressure on the milk from one side of the exit hole to the other, and the force of the surface tension between the milk and the side of the container. It has nothing to do with the Coriolis force as suggested by your correspondent.

The Coriolis force is a real phenomenon. Because the Earth rotates, a fluid that flows along the Earth's surface feels a Coriolis acceleration perpendicular to its velocity. In the northern hemisphere, Coriolis acceleration makes low-pressure storm systems (hurricanes) spin counterclockwise. But in the southern hemisphere storm systems (typhoons) spin clockwise because the direction of the Coriolis acceleration is reversed.

This large-scale meteorological effect leads to the speculation that the small-scale bathtub vortex that you see when you pull the plug from the drain spins one way north of the equator and the other way south of the equator. This is incorrect. The Coriolis force is far too small to have an effect on the direction of bathtub whirlpools or twisting milk coming out of cartons.

The force can be seen in a tub of water only under controlled experimental conditions, including a symmetrical low-friction tub, tight control of thermal currents, and letting the water stand for a long time (a day or more) so the residual fluid motion from filling has ceased.

RAYMOND HALL

Your correspondent's answer to the question is not entirely correct. While it is correct to say that the corkscrew is the end of the whirlpool occurring within the package, he is wrong to suggest that the cause of the whirlpool is the Coriolis effect.

Instead, the "ice-skater" effect is responsible. Any small wobbles you have given the milk package will set the fluid moving inside in one direction or another. As the fluid moves out through the small hole, its angular momentum is conserved. That means that, as it moves into a smaller diameter stream, it spins faster, just as ice-skaters spin faster when they pull their arms in closer. This is also why the corkscrew effect is enhanced with smaller openings.

SONYA LEGG

⚙ Aim and Pour

When I open a carton of milk I have to pour the liquid quickly from the opening so that it goes into my glass. If I tip the carton too slowly, the milk runs down the underside of the carton and pours onto my foot or the floor. Orange juice and other liquids do the same. Why do they stick to the carton when poured slowly?

TOM KHAN

When a carton of liquid is tipped during pouring, the free surface of the liquid in the container is raised relative to the opening. This creates a pressure difference between the free surface and the opening, which forces fluid from the carton. In addition to this pressure force, there are also surface tension forces acting on the fluid that tend to draw the fluid toward the surfaces of the container. At high pouring speeds, the pressure force is much greater than the surface tension forces, and the fluid will leave the carton in an orderly fashion, following a predictably curved (parabolic) path toward a glass below.

However, at low pouring speeds, a point is reached where the surface tension forces are sufficient to divert the path of the fluid jet so that it fails to leave the opening cleanly and becomes attached to the top face of the carton (assuming a flat-topped carton). Once attached to a surface, a jet of liquid will tend to remain attached to that surface, owing to these surface tension forces and a phenomenon known as the Coanda effect. This occurs when a fluid jet on a convex surface (such as a water jet from a tap curving round the back of a spoon) generates internal pressure forces that effectively suck the jet toward the surface.

The combined result of surface tension and the Coanda effect enables an errant flow of fluid to navigate the bend from the top face of the carton around to the carton's side, thus ensuring maximum transport of fluid from the carton to your shoes.

Experiments have shown that when cartons are full, the "glugging" that occurs as air is sucked in to replace the lost fluid causes the fluid jet to oscillate, leading to periodic surface attachment of the jet (and wet shoes) even at relatively high pouring speeds.

BILL CROWTHER
AEROSPACE DIVISION
UNIVERSITY OF MANCHESTER, UK

The Coanda effect or "wall attachment" is named after the Romanian Henri Coanda (1886–1972), who invented a jet aircraft propelled by two combustion chambers, one on either side of the fuselage pointing backward, and situated toward the front of the aircraft. To his horror, on takeoff the jets of flame, instead of remaining straight, clung to the sides of the fuselage all the way to the tail. At least his name has now been immortalized, thanks to this effect.

Some 30 years ago this wall attachment phenomenon was used in machine control systems known as fluidics, where a small jet of fluid was used to persuade the main flow to leave the "wall" to which it was attached and divert to another course. It then became attached to this.

JOHN WORTHINGTON

*A picture of the Coanda, the first true jet aircraft to
be built, in 1910, can be found at www.allstar.fiu.edu/
aero/coanda.htm. The next answer describes a simple
demonstration of the effect—Ed.*

The effect is seen as a general tendency for fluid flows to
wrap around surfaces. An entertaining experiment consists
of taking a vertical cylinder (a bottle of dishwashing liquid
or a wine bottle) and placing a lighted candle on the far side.
When you blow against the bottle, the candle is blown out,
because the current of air wraps around it.

RICHARD HANN

⚙ Double Trouble

*I recently purchased a carton of eggs, each of which was
guaranteed to have two yolks. And the claim was correct.
How does the supplier ensure that each egg has two yolks?*

JOHN CROCKER

These special eggs are a natural phenomenon over which we
have no control. Double yolk eggs are larger than those laid
by the majority of the flock and are set aside to be tested in-
dividually. Demand for double yolkers far outstrips supply,
and we need to be very sure that they do in fact contain two
yolks. Therefore, each egg is checked by holding it against a
bright light. During this process (still known as candling,
from the days when a candle provided the source of light)
the number of yolks will be clearly visible as shadows.

GRAHAM MUIR
STONEGATE FARMERS LIMITED
HAILSHAM, SUSSEX, UK

*Try it at home—you'll be surprised how much of the
inside of an egg you can see.—Ed.*

❖ Frying Problem

When I view the surface of cooking oil in a pan by reflected light, a pattern of honeycomb-like shapes appears as the pan is heated by a gas flame. The unit size of the pattern is smallest where the oil is thinnest. Why is this?

REX WATSON

The honeycomb cells observed in heated cooking oil are known as Rayleigh-Bénard convection cells. At low temperature differences between the bottom and the top of the oil, the heat is dissipated through ordinary thermal transport (collision of individual molecules) and no macroscopic motion can be observed. At greater temperature differences, convection (a collective phenomenon involving many molecules) is a more effective means to transport the thermal energy. The heated cooking oil on the bottom is less dense and wants to rise. The top of the oil cools down by contact with the air and sinks again. This motion becomes circular and creates rolls of liquid, which self-organize into a honeycomb pattern that can be easily observed.

Quite a bit of research has been carried out on this phenomenon—which anyone can create in the kitchen—and we now have an explanation as to why the pattern of the convection cells is honeycomb. The form of the convection rolls depends on the shape of the container in which the liquid is heated. Hexagonal patterns seem to develop easily in round pans. Other containers may lead to long, rectangular rolls, with a square cross section.

As the liquid moves in a circular fashion (up, across, down, and back across), the unit size of the pattern depends linearly on the thickness of the liquid. It is interesting that many variables such as the unit size of the convection cell are determined, whereas the direction of the circular motion is undetermined at the onset of convection. Once a rotation direction (clockwise or counterclockwise) is established, it remains stable.

BERND EGGEN
UNIVERSITY OF EXETER
DEVON, UK

Twenty seconds or so after heat is first applied, the really interesting phase of convection begins suddenly. When the temperature gradient within the oil layer has built up to a certain critical value, each of the many scattered convection currents present in the oil finds that it conserves energy better if it shares its region of descending flow with the downflow regions of its immediate neighbors. This stops any contra-flow problems. This cooperative repositioning of the centers of convection forms a regular pattern of closely packed convection cells. Their honeycomb-like appearance occurs to allow each cell to have the maximum area consistent with sharing its cell walls with its neighbors.

Because of this cell cooperation, convection proceeds vigorously and the rising hot oil can be seen to form a small fountain at the center of each cell. The force that maintains this pattern, in the face of mechanical and thermal disturbance, is the flow of heat energy up through the oil layer. In the same way, a biological system needs energy throughout—in this case food—to maintain its integrity.

A substantial increase in the temperature gradient leads to the breakup of the cell pattern, which may pass through several phases of greater complexity before degenerating into chaos.

ROGER KERSEY

It can be shown analytically that the most efficient flow pattern in a large expanse of fluid transferring heat from bottom to top is hexagonal, with cells about the same width as the depth of the fluid. The hot fluid moves up the center, cools at the surface, and then drops down the perimeter of the hexagon. Similar patterns can be seen on all scales from millimeter-sized experiments to patterns on the surface of the sun.

GARY ODDIE

The readers above have already provided answers to this question. However, as the writer below points out, the previous explanations using the Rayleigh convection model were not wholly correct, for the Rayleigh model applies only if the frying liquid is of sufficient depth—Ed.

The behavior of hot oil in a pan is a classic example of Bénard convection, the unstable motion of fluid on a heated flat plate, which takes the form of regular hexagonal cells of circulating fluid. It is well known that Lord Rayleigh developed a theory to explain this instability. What is not so well known is that his model was wrong.

Rayleigh considered a horizontal layer of liquid with flat surfaces heated from below, and assumed that the instability took the form of parallel, contra-rotating rolls driven by buoyancy forces due to variations in the fluid density. Then, by heuristic arguments, he deduced a size for hexagonal cells close—fortuitously—to that observed by Bénard. He also predicted the minimum temperature gradient across the layer for the onset of this motion, but this was about 100 times greater than the gradient needed to initiate the cellular flow in Bénard's experiments.

Other researchers extended Rayleigh's analysis in various ways. When the flat-upper-surface condition was later relaxed it could be seen that the surface is elevated above rising fluid between adjacent rolls while it is depressed above descending fluid. This is precisely the opposite of what Bénard observed. When Bénard's experiment was repeated it was found that the cells could also be produced when the heating plate was cooled, whereas according to Rayleigh's ideas the fluid should remain at rest. The instability has also been observed for a layer of liquid beneath a plate being heated from above and in space, where gravity and hence buoyancy forces are zero.

In the late 1950s a new model for Bénard convection was developed in which variations of surface tension caused by temperature variations on the surface of the liquid drove the motion. This model also predicted a depressed surface above rising fluid. In reality both Bénard and Rayleigh effects must be present. Conditions determine which predominates. Buoyancy forces drive the motion when there is no free surface or the liquid layer is thicker than about half an inch; otherwise, surface tension governs the flow.

Whichever driving force dominates, it must be sufficient to overcome the effects of viscous drag (which tends to inhibit motion) and diffusion of heat within the fluid (which tends to reduce the temperature gradients) before it can initiate the unstable flow. For buoyancy-driven flows the onset of instability is governed by the Rayleigh number:

buoyancy forces/(viscous drag rate of heat transfer)

while for flows driven by surface tension the corresponding variable is the Marangoni number, in which surface tension forces replace buoyancy forces.

For thin layers the unstable flow takes the form of a regular array of hexagonal cells regardless of the shape of the container. For thicker layers the basic unstable flow is a series of rolls parallel to the container's sides with the direction of flow adjacent to its rim and determined by its temperature relative to its base. These rolls degenerate into polygonal (but not necessarily hexagonal) cells when the temperature gradient is increased.

RICHARD HOLROYD

Stale Tale

Why does a cookie that is left in the open overnight become soft by the morning but a baguette left out for the same length of time become so hard that one could knock someone out with it?

LORNA HALL

Cookies contain much more sugar and salt than baguettes. The finely divided sugar and salt are hygroscopic and soak up moisture from the atmosphere—the osmotic pressure in a sweet cookie is quite high. The dense texture of a cookie helps maintain the moisture by capillary effects.

The baguette, on the other hand, contains little salt or sugar, and has a very open structure. The flour doesn't care if there's moisture around it or not. So, because of their different makeup, one attracts water but the other doesn't. Try a series of different cookies, varying from very sweet, dense ones to light, fluffy sponge cookies. The "overnight sogginess index" increases with density and sugar/salt content. I find that if I put both traditional Italian biscotti (not very sweet and fairly open-textured) and dense, sweet ginger cookies in a sealed container, the biscotti go rock hard and the ginger cookies end up very soft.

CHRIS VERNON

A baguette dries out while a white sugar cookie becomes soft because of the hygroscopicity of the white sugar in the cookie. I researched this last year when entering a competition at the age of 13. We were asked to produce a project about whether cookery was a science.

The water vapor in the air is attracted to the sugar and this makes the cookier softer. Baguettes however, have no sugar in them and therefore have nothing to attract the water vapor, which evaporates to leave the baguette hard.

When we performed the experiment we used three types

of cookies: one made from superfine sugar, another from honey, and the last being a control that had no sweetener. The control lost 0.077 ounce of water after being left outside overnight, and the honey lost 0.072 ounce, but the superfine sugar cookie gained 0.043 ounce. The honey cookie lost water because the atmosphere had a lower concentration of water than the cookie.

TOM WINCH

Starch consists of about 20 percent amylose and 80 percent amylopectin. The key to breads becoming stale is amylose retrogradation. Naturally, loss of moisture is involved or it wouldn't dry out. However, bread can be prevented from losing moisture and still go stale. The linear amylopectin molecules in the starch grains, which are separated by moisture in fresh bread, move closer together and become more ordered as the bread becomes stale, making it stiffer.

The process is temperature-dependent, with the rate fastest at just above freezing and slow below freezing. Studies show that bread stored at 45°F (average fridge temperature) becomes stale at the same rate as bread stored at 86°F. So putting bread in the fridge does not keep it fresher for longer.

ALLIE TAYLOR

The feature referred to in the question has a parallel in legal terms. Here there is a difference between cakes and cookies for VAT purposes. This is important because cakes are subject to VAT, while bread is not. Now we have a new definition: a cookie is something which goes soft when left out, whereas a cake goes hard. What the implications are for VAT on baguettes, I wouldn't like to imagine.

RICHARD BUTLIN

Strings Attached

Why does grilled cheese go stringy?

JOHN MITCHELL

The uncooked cheese contains long-chain protein molecules more or less curled up in a fatty, watery mess. When you heat cheese, the fats and proteins melt; and if you fiddle with the fluid, the chains can get dragged into strings. Grab a bit of the molten cheese and pull, and you get a filament, in much the same way that you can draw and twist cotton batting into yarn.

You can do similar things with polyethylene from plastic bags by heating or stretching the plastic to curl or stretch the long-chain molecules. When the molecules are curled up, the plastic is softish and waxy. When they are stretched into fibers, the result is elastic and strong in the direction of the stretch, although it splits easily between the chains lying along the fiber.

JON RICHFIELD

As the cheese melts, the long-chain protein molecules bind together to form fibers in the liquid mass of melted cheese. I believe that this characteristic can actually be used to measure the protein content of a cheese sample directly. A string of cheese is pulled away from the sample, and the distance to which the fiber will extend away from its attachment point on the main piece of cheese is measured against some reference sample of known protein content.

MIKE PERKIN

Micro Madness

A colleague of mine is in the habit of heating bottled water for his tea in a mug in a microwave oven. When the water is up to the desired temperature he removes the mug.

On several occasions, the water has started to bubble violently after he has added a tea bag. On one occasion, the boiling started when he was removing the mug. It was so violent that it blew 90 percent of the water from the mug—an effect that is obviously quite dangerous. What is happening?

MURRAY CHAPMAN

A portion of the water in the cup is becoming superheated—the liquid temperature is actually slightly above the boiling point, at which it would normally form a gas. In this case, the boiling is hindered by a lack of nucleation sites needed to form the bubbles.

This never occurs in boiling water in a kettle, for example, because the presence of the rough surface of the element, as well as the convective stirring from rising hot water, is sufficient to produce proper boiling. Turbulence in liquids is known to provide enhanced nucleation in other cases: when you pour a cola drink, for example.

In your colleague's case, the addition of a tea bag (and, on the other occasion, simple movement) sufficed to allow bubble formation. Even with a large proportion of the water superheated, only a little will convert to steam, as the amount of latent heat required for this phase change is very large. I imagine that by keeping the cup still and microwaving for a long time, you could blow the entire contents of the cup into the interior of the microwave as soon as you introduced any nucleation sites. It is this sometimes explosive rate of steam production that means you should take great care when using a microwave oven.

RICHARD BARTON

Superheated liquid can boil explosively if something is added, as in the examples given by your previous correspondent, or if the vessel is moved. I have seen a spectacular explosion of a bottle of liquid which had just been removed from a microwave in a laboratory—glass and hot liquid were thrown across the room. This can be avoided by leaving any liquid that has been heated in a microwave to stand for at least a minute before touching it or opening the door. This allows for slight cooling and for the heat to become more evenly distributed. I recommend that everyone does this when heating liquids in a microwave, even to make a cup of tea.

Diane Warne

 ## Green Ham Common

What causes the greenish iridescent sheen that I often notice on bacon and ham? Is it harmful, and why does it vanish when the product is heated? Does this occur on any other foodstuffs?

Georgina Godby

You are likely to find such a sheen on foods containing traces of fat in water. When it is cool this mix separates out microscopically into a film, like oil on a wet road.

In some types of cold meats, such as sliced rump roast or some hams, you may see a handsome opalescence. The beauty of an opal results from light's being refracted and diffracted by arrays of microscopic beads of glassy material in a matrix of a different refractive index. In the meat, the effect is caused by microscopic spheres of fat dispersed in watery muscle tissue. Heat up the meat and you destroy the droplets and change the optical character of the matrix so that the effect is spoiled.

Jon Richfield

The green color that is sometimes observed on bacon and ham is the result of the action of nonpathogenic bacteria that break down the oxygen transport protein myoglobin to produce porphyrin derivatives. These derivatives are large heterocyclic compounds, which can have greenish colors.

STEPHANIE BURTON
DEPARTMENT OF BIOCHEMISTRY AND MICROBIOLOGY
RHODES UNIVERSITY
GRAHAMSTOWN, SOUTH AFRICA

My father, working alone in the Australian bush in the 1920s and 1930s, ate meat either fresh, soon after it was killed, or after it had been hung in a tree long enough for it to turn a brilliant green. The meat was put into a bag to keep the flies off it.

He claimed that the green color showed that the meat was no longer dangerous to consume, and it certainly never killed him. However, there is little doubt that it did change the flavor considerably.

JAN MORTON

Iridescence is caused by light striking a surface and being scattered. The scattered waves interfere to produce a spectrum of colors, which changes depending on the position of the observer. However, if you see a bright green color rather than a mere iridescent sheen, then your meat may be only for the hardy stomachs of those who tramp the Australian bush—Ed.

✿ Floaters

What is the force that drives an isolated and floating piece of wheat or rice breakfast cereal through the milk to the side of the bowl where it aggregates with its companions?

JOHN CHAPMAN

The force is due to an imbalance in the pull from the surface tension of the liquid around the sides of the floating piece of cereal. A simple experiment explains what is happening.

You need tap water and two polystyrene cups plus two small pieces taken from a third cup (two half-inch-diameter circles will do nicely). Fill the first cup to within half an inch of the rim, fill the second cup to the top, and then carefully add more to the second cup until the water is up over the top of the cup but not spilling over, that is until the water is held in a convex bulge above the top of the cup by surface tension.

Now place the small circles of polystyrene in the middle of each. The piece floating in the partially filled cup will, with a little prompting, move to the side of the cup and be held there. By contrast, the piece floating on the convex bulge of the water in the second cup will remain near the center. Furthermore, if you push the piece to the edge of the cup, say with the tip of a pencil, the edge repels the small piece toward the center with considerable force.

This is all caused by the surface tension of the water. In the partially filled cup the water surface curves up to meet the polystyrene. This occurs because water molecules are more attracted to polystyrene than to each other. The water forms the convex bulge at the top of the second cup because the surface tension constrains the liquid surface to the smallest area possible; this effect also accounts for the spherical shape of liquid drops.

The water also curves up to meet the small circle of polystyrene on all sides. Where the water meets the polystyrene of the small circle, the surface tension pulls on each contact point in a down and outward direction provided by the angle of contact with the water. When the circle is in the middle of the cup, the pull on the circle on one side is directly balanced by the pull on the opposite side, because the water curves up to meet the circle equally at all points.

However, if the piece is moved toward the side of the partially filled cup, the upward curve of the water surface near the cup side reduces the curve of the surface in contact with the circle. This increases the outward pull on the side of the circle nearest to the cup edge, resulting in a net force toward the side of the cup.

The effect also accounts for the clumping together of cereal pieces on the surface of milk in your bowl and similar behavior of leaves and twigs on ponds and lakes.

RAY HALL

Maybe it is a defensive strategy: they huddle together like bison, to protect each other from the predator (you). Or maybe it's just the surface tension in the milk.

PER THULIN

The fact that rice and wheat—or any grain, for that matter—can gravitate toward its companions in this fashion, depends upon their being able to "sense" their way toward the common center of mass. This ability is known as the grain of common sense.

Research has shown that when human beings are dropped into a large bowl of milk this flocking or aggregation potential is entirely lacking, thus proving that they don't have a single grain of common sense at all.

MARTIN MILLEN

✿ Vulcanized Eggs

Most substances melt when heated, so why does my scrambled egg turn from liquid to solid as I cook it?

DAVID PHILLIPS

Not all changes between solid and liquid have to do with melting or cooling. Among those that do not are congealing scrambled eggs and polymerization of plastics.

Yolk and albumen—egg white—get their textures from

globular proteins dissolved in them. The globules form because the chain-like protein molecules curl up into balls.
Electric charges at particular positions on the chains hold
the proteins in the shapes suited to their functions. Charges
on the outside of the globules attract water molecules,
thereby repelling other proteins and stopping them from
clumping together.

The balls are not permanent structures, and the charges do
not fasten the proteins very tightly. Rattle them violently—by
heating, for example—and they unravel, exposing their inner
charges. This is called denaturation, because the changed proteins are unsuited to their biological functions. Opposite
charges in neighboring molecules can now meet and stick the
proteins together, congealing them into huge tangles. But your
digestive enzymes can break down such tangles more easily
than the undenatured proteins—so *bon appétit!*

JON RICHFIELD

When you heat a solid, such as ice, you transfer energy to
the molecules, allowing them to break the chemical bonds
that hold them in a solid state. In the liquid state, they have
enough energy to move around, but not quite enough to separate completely from other molecules and form a gas.

When you heat a raw egg, an entirely different process
takes place. The egg is made up of individual proteins floating in water, the proteins consisting of long-chain molecules
twisted and held in a roughly spherical shape by chemical
bonds. As the egg is heated, these bonds break and the molecules unravel, bonding with other molecules to form a network that traps the water and turns the egg solid. Heating
the egg further causes even more bonds to be formed, so the
egg becomes less watery and more rubbery.

NICHOLAS SMITH

Eggs are mainly made of proteins dissolved in water, the
most abundant of which is albumin, constituting most of the
egg white. Proteins are made up of a variety of 20 different

amino acids, which form polymer chains folded densely in a unique and relatively stable 3D structure.

On heating, the egg dehydrates and the protein chains unfold and denature. The heat causes sulfur-hydrogen groups on the amino acid cysteine to oxidize and form covalent bonds between neighboring molecules. These strong, stable bonds are called disulfide bridges, and this cross-linking causes the chains to form networks, so the egg hardens. Disulfide bridges also contribute to the high tensile strength of fingernails and the shape of hair. When hair is "permed," the disulfide bridges are broken by a reducing agent. The hair is then styled into the desired shape and an oxidizing agent is used to reintroduce the covalent bonds and maintain the new shape.

IGNATIUS PANG

⚙ A Matter of Taste

How does temperature affect the taste of food and drink? For example, white wine, tap water, Cointreau, lager, and even chocolate taste much better cold. On the other hand, tea, coffee, and brandy, as well as most cooked meals, taste much better warm or hot. English beer and red wine are better at room or cellar temperature. Why?

ANDREW NEWELL

What we normally refer to as "taste" is more correctly termed flavor, which is made up of taste, irritation, and aroma. Taste per se consists only of the five sensations that can be detected by the tongue: salt, sweet, sour, bitter, and umami. These are not affected by temperature, and neither is irritation from, for instance, chili peppers. But aroma, which is sensed in the nose, is strongly affected by food temperature because it depends on the release of volatile oils. The higher the temperature, the more volatiles are released, and the stronger the aroma and thus the total flavor sensation.

The flavor of foods that have little aroma is enhanced by heating, whereas foods with strong aromas may become overpowering at high temperatures. Red wines, for instance, tend to be drunk at room temperature with meals that have strong flavors, achieving a balance in which food and drink complement each other, rather than canceling each other out. White wines, on the other hand, are often drunk cold with fish or weakly flavored foods. When imbibed at room temperature on its own, however, white wine gives a perfectly pleasant flavor sensation, and one suspects it is just convention for white wine to be served chilled.

Another important effect of temperature on meals is its influence on the viscosity of starch-thickened sauces, which drops at higher temperatures because starches react to heat. The texture of food is very important to people. A meal covered in a cold, starch-thickened sauce is unappealing, but a non-starch-thickened sauce such as mayonnaise covering the same ingredients in a sandwich would be a very different prospect.

There is also a large element of convention and cultural preference involved. We prefer our gazpacho cold but our minestrone piping hot. Beer is served at room temperature in the UK but chilled almost everywhere else. Some people prefer whiskey on the rocks; others—especially in Scotland—find ice an abomination. Hot coffee and iced coffee are equally acceptable to most people, and choice depends mainly on ambient temperature. It's all about circumstance, accompanying flavors, and how we are used to having our food and drink served.

JON F. PRINZ

Lager Doubt

Two advertisements for lager that ran on British TV presented a paradox. The first, for the American brand Bud-

weiser, suggests that the key to good lager is fast shipment from brewery to bottle to drinker. It says fresh lager tastes better. The second, for the Dutch beer Grolsch, makes exactly the opposite claim. It stresses the importance of a long conditioning period to improve flavor before the beer is bottled. Which will produce a better beer and why?

MICK MCCARTHY

As a keen home brewer I feel qualified to answer the question on the aging of lagers.

All true lagers are aged before consumption. "Lager," in fact, comes from a German word meaning to store. After fermentation, the beer undergoes a storage—or lagering—process at low temperature to allow the beer to mature and take on the distinctive clean taste for which lagers are famous. Lagering takes from one week to more than six months, depending on the style. I suspect that both Budweiser and Grolsch undergo this process.

In general, European lagers tend to be more complex than American lagers, which are usually lighter and less intricate in style. Because a complex beer will gain more from lengthy lagering, European lagers tend to be matured for longer than American ones.

After lagering the beer is bottled. Once bottled the beer can spoil easily through exposure to light, oxygen, or high temperatures. Fast shipping and sale minimize the chance of beer spoilage. So, in short, both claims are correct. A lager needs to be matured to develop the correct flavors, and fast shipping, once it is matured, is important.

As for which brand is best, that is a matter of personal taste.

DAVE MARTIN

Both advertisements are correct and the two claims are unconnected.

After fermentation a beer needs first to be matured and aged at a cool temperature, usually between 39 and 45°F.

During this period the residual yeast in the beer continues metabolizing and, because the beer has become nutrient-poor during brewing, reabsorbs compounds that had previously been excreted. The most notable of these is diacetyl, which imparts a butterscotch taste to the beer. Meanwhile the yeast content of the brew steadily falls as the yeast sediments.

Next, the beer is chilled to 30°F or below. This chilling promotes protein coagulation and precipitation, which increase the physical shelf life—or the time the beer takes to go hazy. At the end of all this the beer is filtered and bottled.

From here on it's all downhill. Bottling is traumatic to beer. It is filtered, pumped, packaged, and pasteurized. Some contamination with oxygen is unavoidable, and this immediately gets to work on the compounds in the beer, starting a process of deterioration.

In conclusion, mature it slowly and at length to get a good flavor and then get it into the drinker as fast as possible before it deteriorates. A reasonably good taster can distinguish between a week-old and a month-old bottle from the same batch.

Davɪᴅ Cᴇғᴀɪ

Beer is "raw" immediately after fermentation, and any harsh sugars that are present, such as the Belgian candy used in some beers, burn the nose, while the hops taste like freshly cut grass. The conditioning, or lagering, period is a very slow fermentation during which these raw flavors mellow and the subtler flavors increase in complexity.

At some point the beer reaches its peak of flavor and starts losing taste. A pale ale might peak between one and three months after fermentation, while a high-gravity imperial stout could still be developing years later. Many beer experts think that U.S.-style Budweiser is a very light taste to begin with and that because Budweiser's breweries have strong quality control over every step of their process, they can reduce the need for longer maturation and clarification periods without affecting the taste too much. European lagers, on the other hand, have longer lagering periods because they are far more complex in taste.

After pasteurization, beer is essentially defenseless against degradation. Any temperature swings between the brewery and consumption spoil the taste. Even worse, compounds known as alpha acids from the hops are light-sensitive—photons break down the isohumulones in the liquid, creating 3-methyl-2-butene-1-thiol, which gives the beer a skunky smell and taste. And, yes, it really is the same compound found in skunk spray. Brown bottles slow this process, but clear and green bottles provide almost no protection. Some brewers use chemically modified hop compounds that are resistant to skunking, but even so it is best to use an opaque container, and a steel cask beats anything else.

So both ads are correct. You need a conditioning period for the flavor of the beer to peak, however long that may be. But once you reach that peak you would ideally drink the beer immediately, especially if it is pasteurized.

RON DIPPOLD
BREWER
SAN DIEGO, CALIFORNIA

The time it takes to brew a beer and how quickly it is shipped to consumers are really two different aspects of the overall brewing, packaging, and distribution process. So the two claims are not opposite but complementary in ensuring a good-quality beer.

Anheuser-Busch brews Budweiser for just the right amount of time to give the beer its unique clean, crisp taste. While we assume that all quality brewers understand the time needed to brew and mature their beer, we make the additional effort to bring a freshly packaged beer to the consumer. We even suggest when it should be consumed to ensure it has the freshest taste: within the first 110 days. Our "Born On" date provides this information and recommendation.

We know consumers are looking for the best-tasting beer available, and the fact is, fresh beer tastes better, hence our policy.

ALAN HENDERSON
PRODUCTION DIRECTOR, STAG BREWERY
ANHEUSER-BUSCH COMPANIES, UK

5   Domestic Science

◼ Breaking the Mold

What is the name of the dreaded black mold that colonizes damp places in bathrooms? Because materials produced to remove the mold do not seem to work, nor do household bleaches, detergents, and solvents, can anyone suggest a remedy other than abrasives?

G. W. GREEN

The infamous black mold is the fungus *Aspergillus niger*. The reason it seems so difficult to eradicate is that the visible black manifestation is merely the exposed structure of the fungus, which is mainly composed of the fruiting bodies. In addition to this visible material, there is invariably an insidious network of hyphae or mycelia which lie in the substrate of wallpaper or plaster, and feed on the minerals contained within.

Eradication of the mold requires not only the repeated physical removal of the visible growth but the simultaneous use of a penetrative fungicide capable of permeating the substrate and killing off the unseen root structure. An analogy is trying to eradicate ground elder or horsetail from your vegetable patch by merely trimming the visible plantlets.

ANDREW PHILPOTTS

The Aspergillus fungus has been a constant source of annoyance in public housing projects throughout the nation. It is prevalent where cool, still air deposits condensation next to steel window frames, concrete-screed ceilings, water-tank enclosures, and similar areas.

Current medical opinion is that this fungus is a major source of allergenic disease and that it produces carcinogenic aerosols, so removal of the unsightly nuisance is also important for health.

I have experienced problems when attempting to remove Aspergillus. Table salt and bleach have only limited success, but I finally effected a permanent solution by washing the affected areas several times with a systemic fungicide available from any garden store. This may not, however, be the safest solution, because the fungicide may be as toxic as the fungus.

GLYN DAVIES

The previous answers seem to have fallen into the trap of assuming that any mold that is black is *Aspergillus niger*. In surveys of Scottish housing carried out by my laboratory, the incidence of this species has been rather low.

The most common dark molds in growths on bathroom and other damp walls are likely to be species of Cladosporium, with Aureobasidium, Phoma, and Ulocladium thrown in for good measure. Even green species of Aspergillus and Penicillium can look black when soaked.

The situation is much the same in mainland Europe, so it is likely that it will not be very different in Northumberland or Surrey, unless the bathrooms there come closer than most British bathrooms to providing the subtropical and tropical climates that favor *A. Niger*.

A really black fungus in about 15 percent of houses in Scotland with mold problems is *Stachybotrys aira*. Wallpaper, jute carpet backing, and the cardboard wrapper of gypsum board all provide ideal cellulosic substrates on which it can thrive in very damp conditions. This type of mold may present the greatest hazard of any to the health of occupants of moldy buildings. Its airborne spores are allergenic and powerfully toxigenic. Its toxins inhibit protein synthesis, and are immunosuppressive, an irritant, and hemorrhagic.

It is well known that fodder contaminated by Stachybotrys can kill horses, and it is also harmful to the stable hands. Currently, this mold is of particular concern in North America, where it has been implicated in episodes of building-related illness ranging from chronic fatigue syndrome in adults to fatal pulmonary hemosiderosis in infants. Consequently it has been the subject of lawsuits (one for the sum of $40 million) against builders and employers.

BRIAN FLANNIGAN
DEPARTMENT OF BIOLOGICAL SCIENCES
HERIOT-WATT UNIVERSITY, EDINBURGH, UK

The outside wall of my bathroom in Pimlico used to grow a superb crop of mold which removed the wallpaper and infested the plaster. To remove this Aspergillus, I used a single washing-down with a dark pink solution of potassium permanganate crystals, which effected a cure with no recurrence.

BILL CHRISTIE

Readers should note that potassium permanganate is poisonous if swallowed—Ed.

Household bleach does not remove the staining caused by *Aspergillus niger*. But spraying the surface or painting it with an aqueous solution of 10 percent zinc sulfate prevents the reemergence of the fungus as long as the molecules of zinc sulfate are not washed off.

FARROKH HASSIB

⊗ Hot Stuff

Is it true that hot water placed in a freezer freezes faster than cold water? And if so why does this happen?

Ian Popay

This question was raised many years ago in New Scientist *and never answered satisfactorily. This time we are closer to settling the controversy with answers from several people who have tried the right experiments. Counterintuitive though it may be, it does appear that hot water can freeze more quickly in a refrigerator. Better thermal contact if the water container is placed into an iced-up freezer compartment and a different pattern of convection currents that allow hot water to freeze faster seem the best explanations. Which effect predominates depends on the fridge, the container, and where it is placed—Ed.*

The questioner is correct—it is possible to produce ice cubes more quickly by using initially hot water instead of cold. The effect can be achieved when the container holding the water is placed on a surface of frost or ice. The higher temperature slightly melts the icy surface on which the container rests, greatly improving the thermal contact between the container and the cold surface. The increased rate of heat transfer from the container and contents more than offsets the greater amount of heat that has to be removed. The effect cannot be obtained if the container is suspended or rests on a dry surface.

This effect was first noted by Sir Francis Bacon, who used wooden pails on ice. My own investigation showed that ice cubes could be obtained within 15 minutes rather than 20 minutes if the frost in the refrigerator was deep enough. The incentive to get your ice a little faster is obviously greater in Australia than in cooler countries.

MICHAEL DAVIES
UNIVERSITY OF TASMANIA, AUSTRALIA

But Sir Francis Bacon was not the first to note the effect. Aristotle's account in *Meteorology* implies a similar explanation: "Many people, when they want to cool water quickly, begin by putting it in the sun. So the inhabitants when they encamp on the ice to fish (they cut a hole in the ice and then fish) pour warm water round their rods that it may freeze the quicker; for they use ice like lead to fix the rods."

DAVID EDGE

And it seems untrue that the "effect cannot be obtained if the container is suspended or rests on a dry surface" . . .

This question was raised in *New Scientist* in 1969, by a Tanzanian student named Erasto Mpemba. He discovered that an ice cream mixture froze more quickly when put in the freezer hot than if allowed to cool to room temperature first. I got the same skeptical comments from my teachers as Mpemba did when I based my twelfth-grade project on his question.

First, the project showed that water, either from the tap or distilled, behaved in the same way as the ice cream mixture; the chemical composition is not important. Second, it demonstrated that a reduction in volume by evaporation from hot water was not the cause. Placing thermocouples in the water showed that water at about 50°F reached the freezing point more quickly than water at about 86°F, as predicted by Newton's law of cooling, but that thereafter, water that started off warm solidified more quickly.

In fact, the maximum time taken for water to solidify in the freezer occurred with an initial temperature of about 41°F, and the shortest time at about 95°F. This paradoxical behavior can be explained by a vertical temperature gradient in the water. The rate of heat loss from the upper surface is proportional to the temperature. If the surface can be kept at a higher temperature than the bulk of the liquid, then the rate of heat loss will be greater than from water with the same average temperature, uniformly distributed. If the water is in a tall metal can rather than in a flat dish, the paradoxical effect disappears. We argued that temperature gradients in the tall can were short-circuited by heat conduction through its metal walls.

The question has certainly made me reluctant to take accepted wisdom for granted when it comes to observations which do not fit preconceived notions of what is correct.

J. NEIL CAPE

The classic experiment uses two metal buckets placed in the open air on a cold, preferably windy, night. Stationary water is a poor conductor of heat, and ice forms on the top and around the sides. If the initial temperature is around 50°F, cooling of the core is very slow, particularly as loose ice floats to the top, inhibiting normal convection. There is no means by which the warmer water can come into contact with the cold bucket and transfer its energy to the outside.

If the initial temperature is about 104°F, strong convection is established before any water freezes, and the entire mass cools rapidly and homogeneously. Even though the first ice forms later, complete solidification of the hot water can occur more quickly than if the water starts off cold.

The conditions are critical. Obviously, if the cold bucket starts at 32.2°F and the hot at 211.8°F the experiment is unlikely to cause surprise. The containers must be large enough to sustain convection with a small temperature gradient, but small enough to extract heat quickly from the bucket's surfaces. Forced air cooling on a windy night helps.

It is difficult to generate suitable conditions in a domestic freezer, but the anomaly can be demonstrated in an industrial chiller or a laboratory environmental chamber.

ALAN CALVERD

It's true and I have verified the assertion by experiment. The only limitation is that the container of water must be relatively small so that the capacity of the freezer to conduct away the heat content is not a limiting factor.

Cold water forms its first ice as floating skin, which impedes further convective heat transfer to the surface. Hot water forms ice over the sides and bottom of the container, and the surface remains liquid and relatively hot, allowing radiant heat loss to continue at a higher rate. The large temperature difference drives a vigorous convective circulation which continues to pump heat to the surface, even after most of the water has become frozen.

TOM HERING

This is a cultural myth. Hot water will not freeze faster than cold water in the freezer. However, hot water cooled to room temperature will freeze faster than water that has never been heated. This is because heating causes the water to release dissolved gases (mostly nitrogen and oxygen), which otherwise reduce the rate of ice crystal growth.

TOM TRULL
UNIVERSITY OF TASMANIA, AUSTRALIA

Skeptical Tom Trull from the University of Tasmania might like to stroll over to the refrigerator of the first letter writer, Michael Davies, also from the University of Tasmania. The experimental evidence suggests that the effect is real—the absence of dissolved gas could be another factor that speeds crystal growth.

And there could be yet another factor that none of our letter writers have described—supercooling. More recent research shows that because water may freeze at a variety of temperatures, hot water may begin freezing before cold. But whether it will completely freeze first may be a different matter—Ed.

In scientifically controlled experiments this effect seems to be real. We assume that the temperature in the freezer stays constant during the freezing process, as do the variables of the samples such as container size, conduction, and convection properties inside and outside the container.

However, I feel that one more variable is present, and this is an overlooked temperature variation in the freezer. The temperature oscillation inside the freezer depends on the sensitivity of the thermoelement and the timer of the controller system. We may assume that at the freezer's standard temperature the power used for cooling the freezer operates at a standard rate. If a bucket of cold water is added, it may produce only a small effect on this power output, as it will not trigger the temperature sensor. However, a bucket of hot water may easily activate the sensor and release a short but powerful cooling of the freezer with a cooling overshoot, depending on the timer.

This may be overlooked by an observer at home. I have seen a similar effect in an electric sauna. By fooling the temperature sensor by splashing water I increased the oven's output.

MATTI JARVILEHTO
UNIVERSITY OF OULU, FINLAND

Recent, as yet uncorroborated, research from the University of Washington at St. Louis has offered yet another possibility. Solutes, such as calcium and magnesium bicarbonate, precipitate out if water is heated. These can be seen inside any kettle used to boil hard water. However, water that has not been heated still contains these solutes, and as it is frozen the ice crystals that are forming expel the solutes into the surrounding water. As their concentration increases in the water that has yet to solidify they lower its freez-

ing point like salt sprinkled on a road in winter. This water therefore has to cool further before it freezes. Additionally, because the lowering of the freezing point reduces the temperature difference between the liquid and its surroundings, the heat loss from the water is far less rapid—Ed.

Stick with It

Why doesn't superglue stick to the inside of its tube?

AJIT VESUDEVAN

Superglue will not stick to the inside of its tube because the tube contains oxygen in the form of air but excludes water. Oxygen inhibits whereas water catalyzes.

YVONNE ADAM
BOSTIK LIMITED, LEICESTER, UK

Superglue doesn't stick to the inside of the tube because, being based on a cyanoacrylate monomer, it requires moisture in the form of water or some other active hydrogen-bearing compound to polymerize.

This explains why the best join between two surfaces is made using a thin glue line. An excess thickness of glue will lead to a retarded cure. This moisture sensitivity explains two things: first, why the bottle comes with a seal that's impossible to break without covering oneself in glue; and, second, why the resulting spillage adheres so well to your skin—being warm and moist, skin makes an ideal substrate.

BRIAN GOODLIFFE

The Loctite company in the U.S. discovered the inhibition by oxygen of the otherwise rapid polymerization of cyanoacrylate. That is why the bottle must always be left with plenty of air inside. The liquid monomer converts to solid polymer when oxygen is excluded by trapping the monomer between close-fitting surfaces.

E. BARRACLOUGH

Smell from Hell

Why is it that, whatever they contain, trash cans always smell the same?

RODRI PROTHEROE

The smell is most probably caused by bacteria and fungi feeding on the organic matter in the rubbish. It will be most noticeable if the container is in a warm and damp place.

The smell will not always be exactly the same, but it will be more characteristic of the different organisms than of the type of food they consume. The smell you get from penicillin mold growing on an orange is just the same as that from penicillin mold grown in a laboratory culture. It is pungent, characteristic, and very common.

Analyses of household rubbish have detected very pathogenic bacteria, including *Pasteurella pestis,* the bacterium responsible for causing bubonic plague. So don't sniff too hard.

CARY O'DONNELL

I was pondering this question while taking out rubbish and I realized that garbage cans do not smell the same. A bag containing foodstuffs will invariably be ripped open by local cats unless protected by a trash can, but a bag without food is not. It would seem that, although humans think that the bags smell similar, they are noticeably different to cats.

As for why they smell similar, that would be because they invariably contain similar objects. However, garden refuse, for example, smells nothing like kitchen refuse, which in turn smells nothing like bathroom refuse.

STEWART RAVENHALL

✪ Sticky Problem

Why does sticky tape when pulled from a roll quickly (at half an inch a second) become almost transparent, but when pulled off slowly (at 2 inches a minute) become opaque? Indeed, if while pulling tape off quickly, one pauses for a few seconds, a distinct line is left on the otherwise transparent tape. Can anyone explain?

DAVID HOLLAND

The reason for the difference in behavior lies in the response of the adhesive layer on the tape to the rate at which it is stressed. When the tape is peeled back slowly, the adhesive responds by forming long-drawn-out strands between the two pieces of tape, which break and fall back onto the tape, forming an opaque, rippled surface. These strands can be seen with the naked eye or with a hand lens.

When the tape is pulled off at higher speeds, instead of being able to stretch, the incipient adhesive strands formed break at much lower elongations and produce much less disturbance of the adhesive layer.

The difference arises because of the viscoelastic nature of the polymer that forms the sticky material. The material has a viscous component giving it some of the physical properties of molasses. It also has an elastic component that causes it to behave like a solid material, such as metal in the form of a wire. When molasses is stretched, it forms long strands, almost never breaking, whereas metal wire has a comparatively low elongation and breaks when pulled. At low pulling rates, the adhesive is more like molasses; at high rates, it is more like the metal wire.

Ultimately, the behavior is dependent on the time of relaxation processes at the molecular level. Since time is in some sense equivalent to temperature when we are considering molecular movements, it is interesting to cool the tape in a freezer. Now, pulling at the lower speed produces a much more transparent region. Because there is not enough time for the long-chain molecules to unravel in long strands, the adhesive breaks in a brittle manner.

STEPHEN HANCOCK

Kettle Hums

Why does a kettle sing? Why does the note rise at first, then fade for a while, and then return with a falling frequency.

DON MUNRO
UNIVERSITY OF NEWCASTLE
NEW SOUTH WALES, AUSTRALIA

If you leave the lid off your electric teakettle and switch on, you can see what is happening. The heating element quickly becomes covered with small silvery bubbles, each about one-twentieth of an inch in diameter. These are air bubbles, forced out of solution by heat from the element. Rough parts of the element's metal surface provide nuclei for their growth, and they eventually detach from the hot element and rise to the surface. These bubbles form and burst silently, and are clearly not the cause of the kettle's singing.

After about a minute, the air bubbles are replaced by innumerable smaller bubbles of superheated steam that cling to the growth nuclei on the heating element.

A few seconds later, these primary steam bubbles become unstable. As each bubble forms, its buoyancy tends to pull it away from the hot surface. Being surrounded by water that is still far below the boiling point, the primary steam bubble suddenly condenses, collapsing implosively. Curiously, the bubble does not vanish completely, but leaves behind a tiny secondary bubble, presumably of water vapor, that does not immediately condense but is whirled away by the convection currents. Soon there is such a cloud of these secondary bubbles that the water becomes turbid for half a minute or so.

Meanwhile, the shock waves transmitted through the water by the imploding primary bubbles produce a sizzling sound. You can give this sound a more definite pitch by temporarily replacing the kettle lid. This defines a volume of air above the water surface that resonates to some of the frequencies present in the shock waves.

Soon, the cloud of secondary bubbles clears, and there is a general increase in the size of the primary steam bubbles that are still forming on the element. These are no longer forced to collapse immediately and implosively, since the surrounding water is now practically at the boiling point, so the noise fades away. As they grow, streams of buoyant primary bubbles detach themselves from the surface of the element, condensing in the cooler water half an inch or so above it.

Within seconds, the water becomes hot enough to allow large detached primary bubbles to reach the surface, and now you can hear only the return of sound with the low gurgle of their bursting in the air cavity above the water.

ROGER KERSEY

⚙ Key Problem

Why are the rows on a calculator or number keypad arranged with the lowest numbers at the bottom, when we normally read from the top downward? And why are telephone keypads arranged the other way, with the lowest numbers at the top?

M. D. BERKSON

Mechanical adding machines, based on rotating wheels, always have the 0 button adjacent to the 1 button. By convention, most old adding machines had the numbers increasing in value from the bottom, and this may be a holdover from when the machines had levers on the wheels rather than buttons. When the numbers were put onto a pad arranged as a three-by-three grid with one left over, the order of the numbers, as far as possible, was kept the same.

On a rotary telephone dial the 0 comes adjacent to the 9 because a 0 in the telephone number is signaled by 10 pulses on the line. When the telephones acquired push buttons in a grid the ordering of the buttons was carried over from the old telephone dial.

NICKO VAN SOMEREN

Mirror Image

Why is an image in a mirror inverted left to right but not top to bottom?

KISHOR BHAGWATI

The mirror does not reverse images from left to right; it reverses them from front to back relative to the front of the mirror. Stand facing a mirror. Point to one side. You and your mirror image are pointing in the same direction. Point to the front. Your mirror image is pointing in the direction opposite to you. Point upward. You both point in the same direction. Now stand sideways to the mirror and repeat. You are now pointing in opposite directions when you point sideways. Place the mirror on the floor and stand on it. This time you point in opposite directions when you point upward and your upside-down image points downward. In all cases the direction reverses only when you point toward or away from the mirror.

HILARY JOHNSON

The answer stems from the fact that a reflection is not the same as a rotation. Our bodies have a strong left-right symmetry, and we try to interpret the reflection as a rotation about a central vertical axis. We imagine that the world in front of the mirror has been rotated through 180 degrees around the mirror's vertical axis, and it has arrived behind the mirror where we see the image. Such a rotation would put the head and feet where we expect them, but leave the left and right sides of the body on sides opposite to where they appear in the reflection.

But if instead we imagine the world to have been rotated about a horizontal axis running across the mirror, this would leave you standing on your head, but would keep the left and right sides of your body in the expected positions. The image would then appear top-bottom inverted, but not left-right.

So whether you see the image as left-right inverted or top-bottom inverted—or, for that matter, around any other axis—depends upon which axis you unconsciously (and erroneously) imagine the world has been rotated around.

If you lie on the floor in front of a mirror you can observe both effects at once. The room appears left-right reflected around its vertical axis, while you interpret your body as being left-right reflected around a horizontal axis running from head to foot.

PETER RUSSELL

Actually, a mirror does not invert at all. Look at your face in a mirror: the left side appears on the left and the right on the right.

Now look at someone else's face without a mirror. It has been inverted because of the rotation necessary to turn and look at you: the person's right side is on your left. Other people could equally well turn to look at you by standing on their head, in which case you see their left on your left, but now the top of their head appears at the bottom. We don't normally do this, because it's not very comfortable.

Try this experiment. Write a word on a piece of paper and hold it up to a mirror. You automatically rotate it around a vertical axis and it appears in the mirror inverted left to right. It is this rotation, not the mirror, that inverts the image.

Try the experiment again, and this time when you hold the paper up to the mirror, rotate it around a horizontal axis. The word will be inverted top to bottom.

ALAN HARDING

The problem is caused by the way we visualize the mirror image. We imagine ourselves standing on a carousel, which has done a half turn to put us where we see the image—that is, in the mirror. We see that the top and bottom of our bodies in the mirror image are in the same place, but left and right are reversed.

If instead of a carousel we used a Ferris wheel to rotate ourselves, and imagined ourselves strapped upright in the seat, we would see a different result. When the wheel does a half turn, the mirror image now has left and right in the correct places, but top and bottom are reversed.

The trouble is that we are incorrectly using rotation for these experiments, when, in reality, the mirror reflects front-to-back. Because this is a difficult thing to do with our body, we mentally substitute rotation, which doesn't quite fit what we see.

Generally, we prefer to keep top and bottom correct, so we see a left-to-right reversal in a mirror, although we could see top-to-bottom reversal if we wished.

DAVID SINGER

Sealed in Light

When I am opening some types of self-sealing envelopes I notice that there is a purple fluorescent effect within the gum. This lasts for only a very short time, but can be repeated if I reseal the envelope and pull it apart again. What causes this effect?

STEWARD DUGUID

The colored glow is a form of chemiluminescence. Separating the gummed surfaces requires energy that breaks the attractive forces between the molecules of gum.

Presumably, the act of pulling apart the surfaces supplies excess energy to the gum molecules that lifts them into an excited state. As they decay back to their normal state the energy is released in the form of visible light. The difference in energy between the excited and ground states defines the wavelength and hence the color of the light produced: in this case, purple.

This phenomenon is different from fluorescence, in which light (often ultraviolet) is absorbed and then reemitted at a longer wavelength (in the visible spectrum). Fluorescence gives rise to Day-Glo colors and the blue glow you might observe while drinking tonic water near one of the ultraviolet lamps often found in nightclubs.

PAUL WRIGHT

A similar effect can be seen when you are stripping off a length of electrical insulating tape.

I first noticed this about 30 years ago, and the discovery came, by coincidence, shortly after an explosion in a coal mine. The last people to go down the mine before the explosion had been a crew of electricians.

I wondered if the electricians had been using insulating tape, and so I wrote to the authorities, asking about the possible danger of insulating tape as a source of ignition.

However, I received a reply which stated that the effect was well known, but that there was insufficient energy in the sparks to ignite methane in the mine.

MIKE GAY

I noticed the glow mentioned by your previous correspondents—on Royal Society of Chemistry envelopes in my case—and wondered about ignition of flammable atmospheres. As I facetiously pointed out to the society, members of the Royal Society of Chemistry often open envelopes in environments of lower ignition energy than that of methane. More recently, there has been an explosion attributed to essentially this cause, or at least to peeling off an adhesive label. The entry in future editions of *Bretherick's Handbook of Reactive Chemical Hazards* may run something like this:

Adhesive Labels
Tolson, P., et al., J. Electrost., 1993, 30, 149

A heavy-duty lead-acid battery exploded when an opera-
tor peeled an adhesive label from it. Investigation showed
that this could generate >8 kV potential. Discharge through
the hydrogen/oxygen headspace consequent upon recharging
batteries caused the explosion. The editor has remarked
vivid discharges when opening Royal Society of Chemistry
self-adhesive envelopes.

P. URBEN

Jelly Roll

*Why does a jar containing hand cleaner gel resonate so
when bumped?*

BRUCE BUSWELL

Hand cleaner or degreaser gel, like many materials, has both
a viscous and an elastic side to its nature. It is a gel made up
of a network of weak elastic bonds. These are easily broken
under the shearing action caused by your fingers when you
are using it to clean your hands. If these bonds are not broken down but are subjected to a force within their elastic
limit (such as being bumped in a jar) they will store the energy and oscillate like a spring.

The period of oscillation is related to the bond energy
and length. So when large networks of strong and relatively
short-range bonds are struck, as in a metal anvil, you get
high-pitched ringing tones. Weaker, longer bond networks
like those in hand cleaner gel give low-frequency natural
harmonic oscillations when struck. This low resonance is
quickly damped by the viscous component of the gel, which,
instead of storing the energy of a strike, dissipates it in forms
such as heat and entropy.

WAYNE COLLINS

Hand cleaners or degreasers are either a gel or a very viscous liquid (and there may be a phase change within the normal range of room temperatures). It is somewhat unusual in that most common substances of that nature have high internal friction losses, and you normally get a thud by striking the tin. The low internal losses of the gel suggest that the substance might, on the molecular scale, have some long-range structural order.

Because it is a detergent, its molecules will have an ionic end, which is attracted to water; and a fatty end, which is repelled by water. These molecules could form roughly spherical associations, with the fatty ends outward and water in the center. These would then slide easily over each other when the substance was grossly deformed and could resonate mechanically with low losses if the driving disturbance is of small amplitude. I recall that if some water is added to the gel, the resonating effect is much reduced.

J. M. WOODGATE

Klingoff

Why doesn't plastic wrap cling to a metal bowl as well as it does to an equally smooth glass or ceramic one?

TIM BLOOMFIELD

Plastic wrap works because it acquires an electric charge as it is peeled from the roll. It can then stick to an insulating body by the same mechanism that causes an uncharged piece of paper to stick to the charged glass of your computer or television screen.

For the mechanism to work, the plastic wrap and the object to which it is sticking must be at substantially different electrical potentials. This works when the object is an insulator. When the object is metal, the charge on the film is dissipated throughout the object, negating the effect.

Old plastic wrap taken off the roll doesn't work either.

After a while, the charge breaks away, and the clinginess is lost.

ALISTAIR HAMILTON

Plastic wrap becomes charged with static electricity as it peels from the roll. You can sense the charge by peeling some off and holding it near your face—you will feel the hairs on your cheek stand up. Metal drains away static, whereas glass (or plastic) retains static on its surface. The more static, the greater the cling.

JEFFREY WELLS

Rustle Riddle

What generates the energy that makes plastic shopping bags so noisy?

LUCY BIRKINSHAW

The energy is generated mostly by you, because the bag will not rustle by itself. The noise is caused by sharp movements of the kind you get when a stiff plate buckles or gets rubbed. The bags are made of polyethylene film, which, untreated, is waxy and floppy and not very noisy. It is elastic rather than plastic, so it absorbs stresses quietly.

However, to make the bags, the film is stretched to get it thin enough to be convenient to handle and cheap enough to give away with the purchases. This partly aligns its molecules into stiffer sheets. To make the bags look better and the contents more anonymous, manufacturers add fillers for coloring and further stiffening. The result is a bag that audibly protests every crinkle, crumple, and abrasion.

JON RICHFIELD

◎ First Light

Why do lightbulb filaments usually blow when the light is first switched on, and not at the end of a long evening's use when they are at their hottest and most hardworking?

ALAN STATEN

When a lightbulb is switched on, its delicate filament is hit with a triple whammy.

The resistance of the metal filament rises with its temperature. When it is switched on, its resistance is less than a tenth of its usual working level, so an initial current more than 10 times the rated value surges into the filament, heating it very rapidly and creating thermal stress.

If any part of that filament is slightly thinner than the rest, this area will heat up more quickly. Its resistance per millimeter will be higher than that of the rest of the filament, so more heat will be generated along this stretch than in the adjacent parts of the filament, amplifying the thermal stress.

In addition to all this, the filament is wound into a coil that also acts as an electromagnet. Because of this magnetism, each turn of the coil deflects its neighboring turns so that the initial surge of current jolts the thin, delicate filament, creating mechanical stress.

So it is no wonder the poor thing goes splat when you turn on the light switch.

ROBERT SENIOR

The more electric current that passes through the tungsten-type metal filament of a conventional lightbulb, the more the metal heats up. When the bulb is first switched on, its filament goes very quickly from room temperature to white-hot. This rapid heating exposes the filament to maximum physical and thermal stress. When the current is switched off, the filament is surrounded by the warm structure of the bulb, so it changes temperature more slowly than when it is switched on. The filament therefore has a much greater chance of blowing when switched on than when it is operating or when it is cooling after being switched off.

Ross H. Clements

Light filaments blow when switched on because that is when the highest current and highest temperatures occur. If you measure the resistance of a lightbulb when it is cold you will find it to be far less than the rated resistance.

In a 100-watt bulb I have just measured, the cold resistance was only 6 ohms, but the hot resistance was about 140 ohms. Thus current flow and heat generated are far higher at switch-on than after the filament has heated up to its rated temperature. This is especially true at places where the filament has thinned because of age and evaporation of the metal. The large initial current heats those parts of the filament far above the standard operating temperature, which melts them. A bulb filament works harder at start-up, and heats the thinner sections to far higher temperatures, than during normal operation.

W. Unruh

An incandescent lamp produces light by heating a tungsten wire filament to about 4500°F. At high temperatures tungsten atoms evaporate from the surface of the wire, causing the blackening sometimes seen on the inside of the lamp's glass envelope. This evaporation also causes the filament to gradually get thinner with use.

A filament-destroying hot spot can occur on a filament for two reasons. First, if some turns in the tungsten coil are slightly closer than average, the temperature of these compressed turns will be higher than normal because more of the radiation they emit is trapped. Second, some turns of the coil may be a little thinner than those on either side of them. These turns will have a higher resistance than normal.

Therefore the rate of heat production at hot spots will be higher than in adjacent sections, and the thinner section will also have a reduced surface area, which decreases the rate at which heat can be lost, again contributing to a higher-than-normal temperature.

Because the rate of evaporation increases exponentially with temperature, hot spots will thin faster than cooler sections. Furthermore, as the wire at the hot spot thins, its resistance increases, raising the temperature even more. So the temperature will continue to rise, and wire-thinning will take place at an accelerating rate.

The cold resistance of a filament is about a tenth of that at normal operating temperature. This means that when a lamp is switched on, the initial current is very high compared with the normal running current. If the diameter of the wire at a hot spot has become sufficiently small, the large spike of current at switch-on may melt the wire.

As a small gap forms between the broken ends of the filament, an electrical discharge may cause a spark or arc to form across the gap. This arc may spread to the leads supplying current to the filament. If this happens, the low-resistance arc allows a large surge of current to flow through the lamp, which may in turn cause a fuse to blow or a circuit breaker to trip. The arc may be seen as a flash of light within the lamp.

BILL MADILL
UNIVERSITY OF CENTRAL ENGLAND, BIRMINGHAM, UK

6   Our Planet, Our Universe

 Pole Poll

What time is it at the North Pole?

Nigel Goodwin

There are two answers to this question. The first is that the time for a person is the time determined by her or his circadian rhythm. Initially, this physiological time will be close to the time for the longitude where the individual lived before visiting the pole.

Over a period of weeks at the pole, this time will drift as the individual settles into a rhythm with a period that is usually about 25 hours long.

Of course, there is also a local time, independent of people, unless you are a philosopher residing somewhere other than the pole.

So the second answer is that the time is either daytime (for the six months of summer) or nighttime (for the six months of winter).

I have not been at the pole near the equinoxes, but I would imagine that there are also several weeks of continuous twilight when the sun is just below the horizon.

Will Hopkins

The crux of the question is this: how should a person born and bred at the North Pole, who has never heard of Greenwich or Tokyo or any other place on the Earth, start measuring time?

This can be done as follows. Suppose that it is the dark period of the North Pole when the sun is below the horizon all the time. Fix a horizontal board at the pole and draw a circle on it with two diameters perpendicular to each other. Label the ends of the diameters A, B, C, D around the circle.

At the North Pole, you can see the stars revolving in planes that are parallel to the horizon. The plane of the horizon coincides at the poles with that of the celestial equator.

Then choose some star on the horizon and define 0 hour as the instant when this star passes across the line of sight of point A when viewed from the circle's center (the pole). The crossings of the star across B, C, and D will correspond to 6, 12, and 18 o'clock respectively.

It is then easy to draw other straight lines on the board representing intermediate hours.

If I were required to do this exercise at present (at the North Pole), for reference I would choose the faintest of the three stars in Orion's belt because it lies almost exactly on the celestial equator; it is the brightest of all the stars lying that near, or on, the celestial equator; and it is clearly visible to the naked eye.

The next problem at the pole would be how to decide what time it was in the summer, when no stars can be seen because it is constantly daylight.

Having drawn the hour lines in the winter, you must wait for the sun to appear above the horizon. At the moment it is sighted at the approach of the Arctic spring, we make a note of its direction on the board. The hour line on which it falls can be called the time of sunrise in the 24-hour system that was devised during the winter.

The sun will revolve, like the stars in winter, in a plane parallel to the horizon; but unlike our reference star, which always revolves in the same plane, the sun's plane will be higher up day by day, ultimately reaching a highest level of 23.5 degrees from the horizon.

Then it will become lower and lower again until, six months after our first sighting, the sun will disappear below the horizon.

D. S. PARANSIS
LULEÅ UNIVERSITY OF TECHNOLOGY
SWEDEN

This isn't a sensible question: time is independent of location. When it is 1800 GMT in London, it is also 1800 GMT at the North Pole, in Timbuktu, or on the far side of the moon.

One could ask: "What time zone is it at the North Pole?" but this also fails. Time zones are defined politically and administratively rather than by geography. Because the North Pole is floating on the high seas no time zone is defined for it.

Attempts to define a natural time astronomically also fail. Noon is when the sun is due south, but at the North Pole the sun is always due south. Noon is when the sun is at its highest, but the height of the sun is essentially constant at the North Pole. Noon is halfway through the period of daylight, but at the North Pole it is light for six months and then dark for six months.

MIKE GUY

Time, from a geophysical point of view, is related to the position of the sun over the Earth and to the position of the observer. Because any direction from the North Pole is south, the sun is always in the south; and whatever the time is at the North Pole, it is always the same time.

What time is that? The international date line runs through the North Pole, leaving the pole sitting eternally between one date and the next. In other words, it is always midnight at the North Pole.

This, of course, explains how Father Christmas manages to deliver presents to every good little boy and girl throughout the world in a single night.

He just heads out of his grotto due south (which from the North Pole is any direction), drops off as many presents as he can fit on his sleigh, and then heads back home—where it is exactly the same time as when he left. So he can then drop off more presents, return home, and so forth.

PATRICK WHITTAKER

The North Pole is, of course, the true spiritual home of the politician because in answer to the question "what time is it?," she or he can, with all honesty, say "what time do you want it to be?"

PAUL BIRCHALL

Deep Breath

Is it true that every time we take a breath of air or swallow a mouthful of water, we consume some of the atoms breathed or swallowed by Leonardo da Vinci (as I read in a children's science book in 1960)?

STEVE MOLINE

We do indeed breathe in a considerable number of molecules that once passed through Leonardo's lungs and, unfortunately, Adolf Hitler's or anyone else's for that matter. The calculation is not too difficult and is as follows.

The total mass of the Earth's atmosphere is about 5×10^{21} grams. If we take air to be a mixture of about four molecules of nitrogen to one of oxygen, the mass of 1 mole of air will be about 28.8 grams. One mole of any substance contains about 6×10^{21} molecules. So there are about 1.04×10^{44} molecules in the Earth's atmosphere.

A single mole of any gas at body temperature and atmospheric pressure has a volume of about 6.7 gallons. The volume of air breathed in or out in the average human breath is about 1 quart. So we can assume that Leonardo da Vinci, in one breath, breathed out about 2.4×10^{22} molecules.

The average human takes, say, 25 breaths per minute, so during his 45-year lifetime (1452 to 1519) Leonardo would have breathed out about 2.1 x 10^{31} molecules. So, about 1 molecule in every 5 x 10^{12} molecules in the atmosphere was breathed out by Leonardo da Vinci.

However, because we breathe in about 2.4 x 10^{22} molecules with each breath, there is a pretty good chance we breathe in about 4.3 x 10^{9} molecules that Leonardo breathed out. In fact, you can also show in a similar way that you probably breathe in about 5 molecules that he breathed out in his dying gasp.

Of course, there are some pretty crude assumptions involved here in order to arrive at the conclusion. We assume that there has been a good mixing of Leonardo's molecules with the rest of the atmosphere (quite likely in 500 years), that he didn't recycle some of his own molecules, and that there is no loss from the atmosphere due to later users, combustion, nitrogen fixation, and so on. But there is still scope for a considerable loss of molecules without its affecting the main point of the calculation.

When we know that the number of molecules in the hydrosphere is 5.7 x 10^{46}, we can make similar calculations for water. These show that a mouthful of liquid contains about 18 x 10^{6} molecules that passed through Leonardo during his lifetime. So, in addition to breathing in his breath, there is also a pretty good chance of picking up some of Leonardo's urine in every glass of water that you drink.

PETER BORROWS

The law of conservation of matter ensures that atoms are constantly being recycled in the universe. Gravity ensures that most of those on the Earth stay there. Some of the atoms floating around were breathed by Leonardo, although the number of these atoms compared with all those in the Earth's atmosphere would make them few and far between.

However, considering the length of time in which, say, the dinosaurs inhabited the Earth, you can be pretty sure that every breath you take contains what was once part of one or more of these creatures, and that every apple you eat has many atoms that were once part of an animal, even a human. Of course all this could have some very worrying implications for vegetarians.

GLENN ALEXANDER

This question provides some food for thought for homeopaths. There is a very high probability that a cup of water contains a few homeopathic molecules that are effective in countering every illness we may have, and at no cost.

LASSI HYVARINEN

◈ Midday Madness

There are more hours of daylight after noon than before it, particularly in summer. Does this mean midday is in the wrong place?

DEAN SHERWIN

"Midday" refers to the moment when the sun crosses the local meridian, which is one of the imaginary lines joining the North and South Poles at 90° to the equator. If you set your watch so that noon occurs as the sun crosses the meridian, the day will be of equal length on both sides of noon.

However, this system would mean that you would need to reset your watch if you traveled only a short distance east or west. To avoid the confusion this would cause we use time zones—areas throughout which we say that the time is the same, irrespective of the actual meridian. Time zones are nominally 15° wide but in practice they vary in size and shape because of political, geographical, and practical considerations. The difference between the local meridian and the time setting of your time zone can be quite apparent if you live near the edge of an oddly shaped time zone.

DAVID EDDY

The development of time zones is usually attributed to the development of the railroad system in the United States. That system runs predominantly east to west. Until the advent of the railroads, most towns followed local time and the clock midday was well aligned with the solar midday. Then trains began to travel from town to town so quickly that the continual compensation for different local times caused timetable difficulties, prompting the development of time zones.

KEITH ANDERSON

The standard time used in Britain is based upon the Greenwich meridian and the latitude of your correspondent Dean Sherwin—in Reading, UK—is almost the same as that of Greenwich, but his longitude is 1° west. Sunrise, local noon, and sunset therefore occur about 4 minutes later than at Greenwich, and Reading's local time is actually 4 minutes later than the standard clock time used in Britain.

This means that, at Reading, the duration of daylight after noon, as shown on the clock, is on average longer than the duration of daylight before noon. East of the Greenwich meridian, afternoon daylight is, on average, shorter than morning daylight. At Greenwich the difference between the duration of morning and afternoon daylight, averaged over a year, is zero.

On any particular day, the difference between the duration of morning daylight and afternoon daylight depends not only upon the latitude and longitude of a place but also upon the equation of time. This is the difference, in time, between the mean sun, which gives us clock time, and the true sun. It is caused by the eccentricity of the Earth's orbit around the sun, and the tilt of the Earth's axis in relation to its orbital plane. The equation of time varies during the year from minus 14 minutes to plus 16 minutes, and it is the main reason for the difference between the time that you will calculate by looking at a sundial and that shown on a clock. There is also a slight difference between morning and afternoon caused by that day's portion of the sun's annual movement around the ecliptic.

The combination of the above effects can create a difference between morning and afternoon daylight of more than half an hour at Reading.

None of this means, however, that midday is in the wrong place. It means merely that the standard time system, whose simplicity and uniformity are essential for communication, is necessarily an approximation of the sun's complex apparent motion.

The further lengthening of the afternoon daylight, and shortening of before-noon daylight, during the months of British Summer Time are of course, the intended result of the forward movement of clocks by one hour.

DAVID LE CONTE
ASTRONOMICAL SOCIETY OF GUERNSEY

Midday Greenwich Mean Time is only the middle of the day at the Greenwich meridian. If you are west of Greenwich, as Reading is, the sun rises later and sets later, so 1200 GMT will be earlier than the midpoint between sunrise and sunset. The sun travels 360° in 24 hours, or 15° in one hour. Hence, as I write this in north London (0° 10' west), 1200 GMT is 24 seconds before midday, but if I lived in Swansea (3° 56' west), 1200 GMT would be nearly 16 minutes before midday.

If we use Central European Winter Time, 1200 is 6 minutes before midday in Berlin (13° 30' east), but it is more than 50 minutes before midday in Paris (2° 15' east).

The most extreme example is Lisbon in Portugal (9° west), which has recently adopted Central European Time—during summer, 1200 is 2½ hours before midday.

NIGEL WHEATLEY

◈ Clear Day Blues

Why (on a clear day) is the sky blue?

JASPAR GRAHAM-JONES

The sky is blue because of a process called Rayleigh scattering. Light arriving from the sun hits the molecules in the air and is scattered in all directions. The amount of scattering depends dramatically on the frequency, that is, the color of the light. Blue light, which has a high frequency, is scattered ten times more than red light, which has a lower frequency. So the "background" scattered light we see in the sky is blue.

This same process also explains the beautiful red colors at sunset. When the sun is low on the horizon, its light has to pass through a large amount of atmosphere on its way to us. During the trip, blue light is scattered away, but red light, which is less susceptible to scattering, can continue on its direct path to our eyes.

RICK ERAHO

The sky is blue because of a process known as Rayleigh scattering. According to classical physics, an accelerated charge emits electromagnetic radiation. Conversely, electromagnetic radiation may interact with charged particles, causing them to oscillate. An oscillating charge is continually being accelerated and hence will reemit radiation. We say that it becomes a secondary source of radiation. This effect is known as the scattering of the incident radiation.

The atmosphere is, of course, composed of various gases that together form air. We may treat each air molecule as an electron oscillator. The electron charge distribution of each molecule presents a scattering cross section to the incident radiation. This is essentially an area upon which the incident radiation must fall for scattering to occur. The amount of scattered radiation will depend upon the magnitude of this cross section. In Rayleigh scattering the cross section is proportional to the fourth power of the frequency of the incident radiation. Sunlight is composed of various visible frequencies ranging from low-frequency (red) to higher-frequency (blue) light. Because it is of a higher frequency than other visible components, the blue part of the sun's spectrum will be scattered more strongly. It is this scattered light that we see, and so the sky appears to be blue.

Incidentally we are also able to explain why sunsets are red. When the sun is close to the horizon its light must travel through more atmosphere. The blue light will be scattered strongly, whereas red light, because it is of lower frequency, is less prone to scattering and so is able to travel straight to the observer.

D. ROBERTS
PHYSICS DEPARTMENT
UNIVERSITY OF SHEFFIELD, SOUTH YORKSHIRE, UK

 # Chinese Puzzle

The Great Wall of China is commonly cited as being the only nonnatural object visible from space. For an object to be visible when viewed from above, the eye must be able to resolve it in two dimensions. The Great Wall of China is immensely long, but very narrow. If the eye is able to resolve its width from space, many other objects, such as the Great Pyramid of Cheops, should be large enough to be seen in two dimensions, despite having a much smaller total area. Is the ability of the eye to resolve objects in the smaller of the two dimensions affected by the magnitude of the larger (and if so why), or is the claim made for the Great Wall of China incorrect?

A. R. MacDiarmaid-Gordon

The claim is incorrect. It is well known as one of the most widely believed urban legends, perhaps second only to the famous one about mass suicide by lemmings.

A person with perfect eyesight is able to resolve up to about 1 minute of arc without binoculars or a telescope. The Great Wall of China is, very approximately, 20 feet wide. This means that it is not directly visible above an altitude of about 13 miles, or just over twice the height of Mount Everest. Even if its shadow is taken into account, this would make it visible, in places, only up to perhaps about 40 miles at the most. Because of atmospheric drag, this is still below the height necessary for a stable spacecraft orbit.

There are, however, many man-made objects which are visible from outer space, the largest being the Dutch polders, or reclaimed land. Cities too can be seen at night because of the bright lights.

D. Fisk

It is well known that the human eye can pick out long objects much more easily than short ones, so the Great Wall of China is certainly a candidate for being visible from the moon. However, the wall is, in places, a broken-down edifice and is often scarcely visible on the ground, never mind from space. H. J. P. Arnold, photographic expert and skilled astronomer, has studied this problem and concludes that seeing the wall from the moon is a physical impossibility.

Neil Armstrong of Apollo 11 has stated that the wall is definitely not visible from the moon. His fellow astronaut Jim Lovell of Apollo 8 and 13 made very careful observations and says that the claim is absurd. Jim Irwin of Apollo 15 has said that seeing the wall is out of the question.

Photographs from uncrewed probes do show that the route of the wall is sometimes shown by sand that is blown onto the windward side, but that the wall itself is not visible. The end, perhaps, of yet another legend.

Robert Brown

⚙ Mad Tidings

Can anyone explain in simple and commonsense terms why there is a high tide simultaneously on both sides of the Earth?

Pat Sheil

In considering the origin of tides we must disregard the Earth's daily rotation around its axis and concentrate only on the revolution of the Earth-moon system.

This revolution takes place around the system's common center of gravity, which is about halfway from the surface to the center of the Earth, and causes every point in the Earth's interior or on its surface to describe a circle of radius equal to the distance of the common center of gravity from the Earth's center.

Therefore, at every point there is a centrifugal force of the same magnitude and in the same direction: away from the moon, parallel to the line joining the Earth-moon centers. This centrifugal force is distinct from the one caused by the Earth's rotation, which we are disregarding.

Every point of the Earth also experiences a gravitational force as it is pulled toward the moon, the direction of this force being different for different points of the Earth.

The resultant of these two forces creates the tide-generating force. If we now consider two points on the Earth's surface, one directly below the moon and the other on the far side, it turns out that the moon's gravitational force at the near point is greater than the centrifugal force, which, as we have seen, is away from the moon.

The far point is farther away from the moon by one Earth diameter, and the moon's gravitational force there happens to be smaller than the centrifugal force, so the net force on water at the far point is away from the moon.

In most popular accounts, the simultaneous occurrence of tides at the two opposite points is explained by asserting that while the moon pulls the water at the near point some distance, it pulls the Earth's body a little less.

But this explanation does not clarify why a system like that will not simply collapse under the mutual gravitational attraction between the Earth and the moon.

D. S. Parasnis
Department of Geophysics
Luleå University of Technology, Sweden

Ignoring the effects of other bodies, we can say that the center of mass of the Earth and the center of mass of the moon are both in free fall, following orbits around the common center of mass of the Earth-moon system, which gravity and centrifugal acceleration precisely cancel out.

Over most of the Earth's surface, though, this cancellation is not precise, because you're either nearer to, or farther away from, the moon, but still forced to orbit at the same rate as the Earth's center of mass.

For the ocean on the side of the Earth facing the moon, lunar gravity dominates centrifugal force, so water bulges toward the moon.

On the opposite side, centrifugal force dominates, so water bulges away from the moon. Both bulges produce high tides.

In effect, sea level—which would otherwise be spherical—is stretched along the Earth-moon axis into an ellipsoid, and as any point on the Earth rotates into and out of either bulge, the local tides flow, then ebb.

GREG EGAN

The simultaneous high tides on opposite sides of the Earth are a result of an imbalance between gravitational forces and centrifugal forces. Tides are caused by the gravitational interaction of Earth and moon, and to a lesser extent the Earth-sun interaction.

Although we think of the moon as orbiting the Earth, in fact the moon and the Earth both orbit their common center of mass, which is close to, but not exactly at, the center of the Earth. The centrifugal forces generated by the orbital motions of each body just balance the gravitational pull of the other body.

However, the balance is exact only at the center of each body. On the side of the Earth nearest the moon, the moon's gravitational pull is slightly greater and the centrifugal force slightly less than at the Earth's center, so water here is pulled out toward the moon. On the opposite side of the Earth, the gravitational pull is slightly less and the centrifugal force slightly greater, so here water is thrown outward away from the moon.

MARK BERTINAT

⚙ Paddle Puzzle

Glenbrook Infants School went to the seaside for our sum-mer outing. We had a nice time, but please can you tell me why the sea is salty. My mother doesn't know.

JOHN CONNOLLY

The sea is salty because the rivers that flow into it wash salts and other minerals out of the ground. The salts dissolve in the rivers and the rivers flow into the sea. As the sun evaporates the water from the sea to make clouds, it leaves the salts and minerals behind, so the sea is saltier than rivers and lakes.

JACK CAVE-LYNCH (AGE 9)

John Connolly is a brainy guy
Asking questions and wondering why
The salty sea which is such fun
When splashing in the waves and sun
Is not freshwater from the tap
Or from a bottle with a cap;
So he will learn that salt and sea
Mix just like sugar into tea
And that many other kinds of salt
Dissolve into this briny malt,
Sodium chloride, the salt of table
Has other friends within its stable
Potassium Ch, magnesium Ch, and iodide
All flow solvent with the tide.
So now, dear John, you clever lad
Off you go—tell mom and dad!

RAY HEATON

Energy Loss

What is the so-called "slingshot effect" used to accelerate interplanetary spacecraft? It obviously makes use of the gravitational attraction of a planet, but my naive understanding of physics tells me that any kinetic energy gained on approaching a body would be lost as potential energy on leaving. How does the spacecraft extract energy from the planet?

DAVID BATES

I had the same problem as your questioner when I first heard about Voyager using the "slingshot effect." Clearly, a probe will not make any net gain in energy by simply falling through a stationary gravity field.

However, Jupiter and its gravity field are moving around the sun at a speed of about 4,200 feet per second, and any probe passing behind the planet will be accelerated by this moving gravity field much as a surfer is pushed forward by a wave. The energy comes not from the gravitational field but from the kinetic energy of the moving planet, which is slowed by the tiniest amount in its orbit, causing it to drop ever so slightly closer to the sun.

The planet speeds up as it falls toward the sun, and paradoxically it ends up moving more quickly than it did before. Moving Jupiter closer to the sun by 10–15 feet (about the diameter of a proton) would yield more than 416 megajoules.

MIKE BROWN

The Living Dead

In one of her songs, the American artist Laurie Anderson uses the refrain "Now that the living outnumber the dead . . ." Is this true? If so, when did it happen? If not, when might it happen, if ever? Do we have good estimates of population numbers before recorded history?

JOHN WOODLEY

The answer below is based on some calculations that were published by the International Statistical Institute.

If the world population had always been increasing at its present rate, doubling within an average human life span, then the living would indeed outnumber the dead.

However, this is not what has happened. There have been very long periods in the past when the population hardly grew at all, but when deaths continued to accumulate. For historical periods, there is a surprising amount of information on population figures, including censuses conducted by both the Romans and the Chinese.

Before then, there are estimates based on the area of the world that was under cultivation or used for hunting, and of the numbers of people who could be supported per acre using these methods of food production. According to estimates assembled by J-N. Biraben, the world population was about 500,000 in 40,000 BC. It grew—but not at a steady rate—to between 200 million and 300 million in the first millennium after Christ, and reached 1 billion early in the nineteenth century.

On multiplying the population numbers by the estimated death rates, you discover that the total number of deaths between 40,000 BC and the present comes to something in the order of 60 billion. The present world population is still only about 6 billion.

Although no great claim can be made for the accuracy of the historical estimates, the errors can hardly be so large as to affect the conclusion that the living are far outnumbered by the dead. This has always been the case, and will continue to be so into the indefinite future.

ROGER THATCHER

In the Garden of Eden, the living (2) outnumbered the dead (0).

G. L. PAPAGEORGIOU

In the Indian epic poem *Mahabharata* the eldest Pandava, Yudisthira, was asked many questions, including the one posed above, by the god Yama, who was the keeper of the underworld and all that is righteous, to test Yudisthira's knowledge, power of reasoning, and truthfulness.

Yama was disguised as a stork guarding a pond from which Yudisthira's four brothers drank without being able to answer a single question and were all struck dead. The stork Yama asked, "Who are the more numerous, the living or the dead?" Yudisthira answered, "The living, because the dead are no more!"

Yama accepted this and all the other answers given by Yudisthira and, with great pleasure because Yudisthira was actually the son of Yama, blessed him and revived all of his dead brothers.

SHAFI AHMED

Snow Laughing Matter

Would it be possible to reduce the greenhouse effect by painting roofs of buildings white to reflect sunlight in the same way the polar icecaps do? Does a paint exist that would mimic the reflective properties of snow?

PAUL NOLAN

Painting roofs white would reflect more sunlight and it might also compensate for global warming. The Global Rural Urban Mapping Project (GRUMP), undertaken by the Earth Institute at Columbia University in New York, shows that roughly 3 percent of the Earth's land surface is covered with buildings.

The Earth has an albedo of 0.29, meaning that it reflects

29 percent of the sunlight that falls upon it. With an albedo of 0.1, towns absorb more sunlight than the global average. Painting all roofs white could nudge the Earth's albedo from 0.29 toward 0.30. According to a very simple "zero-dimensional" model of the Earth, this would lead to a drop in global temperature of up to 2°F, almost exactly canceling out the global warming that has taken place since the start of the industrial revolution. A zero-dimensional model, however, excludes the atmosphere and, crucially, the role of clouds. It would be interesting to see if more sophisticated models predict a similar magnitude of cooling.

MIKE FOLLOWS

A better use of roofs would be to make them mini power stations by installing photovoltaic tiles. This would displace a significant proportion of the fossil carbon that we emit without relying on perturbing the Earth's delicate and complex climate system. Sure prevention is much better than uncertain cure.

MIKE HULME

⚙ Fading Star

As the sun produces energy, it presumably loses mass and its gravity weakens. Are the planets slowly spiraling outward? If so, by how much, and by the time the sun becomes a red giant, how far out would the Earth be?

MIKE GANLEY

The sun loses more than 4 million tons per second: the mass equivalent of the energy it produces through thermonuclear reactions. Another few million tons are lost in solar wind and other particle emissions. However, even after 2 billion years, this loss constitutes only one ten-thousandth of the sun's mass. So the change in the Earth's distance from the sun will be of the same fractional order.

The situation will be more drastic when the sun eventually becomes a red giant, 6 billion years or so from now. Then, the solar radius will be 100 times its present value. Some of the latest estimates suggest that in its giant stage the sun may well engulf Mercury, Venus, and Earth, while planets more distant than Mars will survive and continue to orbit the sun when it later becomes a white dwarf.

If we assume the final mass of the sun at the white-dwarf stage to be 0.6 of its present value, the dimensions of the planetary orbits in the very far future will be about 80 percent greater than they are now, for the reasons your questioner suggests.

C. SIVARAM
INDIAN INSTITUTE OF ASTROPHYSICS
KORAMANGALA, BANGALORE, INDIA

Amazingly, although the sun converts more than 4 million tons of its mass into pure energy every second and will continue to burn hydrogen until it becomes a red giant several billion years from now, it will even then have lost only a tiny proportion of its present mass. In order for the Earth to conserve its angular momentum, the radius of its orbit will have to increase at the rate of only about a half an inch a year.

However, this will not be sufficient to compensate for the steadily increasing luminosity of the sun. So Earth is destined to follow in the footsteps of its celestial companion, Venus, and undergo a natural runaway greenhouse effect—unless human activity contrives to accomplish it rather sooner.

MIKE FOLLOWS

7　Weird Weather

⚙ Forked Frolics

Why does lightning fork and what is the diameter of a bolt of lightning?

MICHAEL LEE

Lightning usually brings the negative charge from a thunderstorm down to the ground. A negatively charged leader precedes the visible lightning, moving downward below the clouds and through air containing pockets of positive charge. These are caused by point discharge ions released from the ground by the thunderstorm's high electric field.

The leader branches in its attempt to find the path of least resistance. When one of these branches gets close to the ground, the negative charges attract positive ions from pointed objects, such as grass and trees, to form a conducting path between cloud and ground. The negative charges then drain to ground starting from the bottom of the leader channel. This is the visible "return stroke" whose luminosity travels upward as the charges move down. Those branches of the leader that were not successful in reaching the ground become brighter when their charges drain into the main channel.

Photographs of lightning often overestimate the channel width because the film can be overexposed. Damaged objects that have been struck by lightning show channel diameters of between one-tenth of an inch and 4 inches.

R. SAUNDERS
ATMOSPHERIC PHYSICS GROUP
MANCHESTER UNIVERSITY, UK

Wave Power

What mechanism transforms gusting wind energy into the regular wave train of ocean swells, and what determines their amplitude and frequency?

FRANK SCAHILL

When the wind blows over a flat sea surface, small ripples form. These probably correspond to individual strong gusts, are disorganized, and have no fixed direction or frequency.

However, as the wind continues to blow, two things happen. First, the waves interact with each other to produce longer waves that have lower frequency. Second, the wind pushes these larger waves and puts even more energy into them. As long as the storm lasts, the wind will make the waves larger and the wave dynamics will create longer and longer waves.

Some waves will become too steep and break, but in general the total amount of energy will keep increasing. These locally generated waves are known as "wind-sea." Their energy depends on how long the wind has been blowing (its duration) and over what distance (the fetch). The waves on the sea surface are not a simple wave train but a complicated random surface.

It is impossible to give a simple amplitude and frequency for a system as complex as this. Instead, significant wave height, the mean height of the highest third of the waves, is used to describe how large the waves are; and the peak period, the time between the dominant or most energetic waves, is used as a measure of frequency. On average, there will be a wave twice the significant wave height every 3 hours.

Eventually, the energy put into the sea by the wind will be

balanced by the loss of energy, mainly through waves breaking. At this point, the waves will cease to grow and the sea is described as "fully developed." In a wind of 65 feet per second (a Force 8 gale), a fully developed sea would have a significant wave height of 30 feet and a peak period of 15 seconds.

Waves can travel thousands of miles from the point of generation. Unlike light or sound waves, sea waves travel faster as they become longer (and as the frequency gets smaller).

Waves that escape from the storm that generated them are known as "swell." They have a much narrower range of periods and are almost regular wave trains. Because no more energy is put into them, none is dissipated by breaking, and they continue across the ocean until they hit the land.

Because different frequencies travel at different speeds, as swell travels across the ocean it separates into its individual components. So the significant wave height and peak period of the swell are set by the wind speed, duration, and fetch from the storm that generated them.

Peter Challenor
Southampton Oceanography Centre
Hampshire, UK

Wind energy first gives rise to a wind-sea. Waves in a wind-sea are steeper and more chaotic than swell, and are accompanied by whitecaps, the breaking crests of waves. The longer the wind blows, the longer the wavelength of the predominant waves in the wind-sea.

When the wind ceases or the wind-sea waves move out of the generating region, whitecapping continues for a time and is accompanied by a lengthening of the waves, until they are no longer steep enough to sustain whitecaps. The wind-sea then becomes swell.

Surface waves on liquids are dispersive: this means that different wavelengths travel with different velocities. The longer wavelength swell travels faster and arrives at the observer first.

With the passage of time, the swell wavelength becomes shorter as shorter, slower wavelengths arrive. Swell from a storm that formed thousands of miles away may persist for several days, steadily getting shorter because of its dispersion.

Dispersion acts as a filter, so only swell within a narrow bandwidth is present in one region of ocean at any time. This is why swell looks so uniform when reviewed from an aircraft.

Generally, swell decreases in amplitude as it travels away from the source region because its energy is spread over an ever-larger region of ocean.

However, this is not the whole story. A following wind will generate a wind-sea that can transfer some of its energy to the swell and increase the amplitude of the swell without changing its wavelength. Likewise, an opposing wind-sea can diminish a swell.

JOHN REID
FORMERLY OF HOBART LABORATORIES OF THE DIVISION OF
MARINE RESEARCH
TASMANIA, AUSTRALIA

❖ Clouding the Issue

Why do clouds darken to a very deep gray just before rain is about to begin or just before a heavy thunderstorm?

MATT BOURKE

Clouds darken from a pleasant fluffy white just before rain begins to fall because they absorb more light.

Clouds normally appear white when the light that strikes them is scattered by the small ice or water particles from which they are composed. However, when the size of these ice and water particles increases—as it does just before clouds begin to deposit rain—this scattering of light is increasingly replaced by absorption.

As a result, much less light reaches the observer on the ground below and the clouds look darker.

KEITH APPLEYARD

❖ Tainted Tint

I have a photochronic coating on my glasses. Under a blazing Caribbean sun they were only moderately tinted. However, under a weak midwinter sun in the United Kingdom they go almost black. Why?

JEFF LANDER

We have two types of explanation here: one physical, one chemical. It seems likely that chemistry is responsible for the greater effect—Ed.

I can only assume the questioner was walking around in the Caribbean, rather than lying on his back getting a tan. If so, the following may explain his experience.

The sun would be fairly low in the British winter sky, its rays shining almost directly on, and perpendicular to, the vertical plane of his lenses. In the tropics, the sun could be almost directly overhead, and if he was walking around, the sun's rays would strike his glasses edge-on. A sliver of radiant energy would be all that each lens would receive, thus reducing their shading reaction.

CHARLES KLUEPFEL

One of the little details opticians fail to mention about photochromic glasses is that they do not work as well when hot. Particles of silver halide trapped inside the glass are normally transparent, but when struck by ultraviolet light, they disassociate into halogen and metallic silver, which darkens the lenses.

As both components are trapped inside the glass, they will recombine when UV light is removed—when you go indoors—becoming transparent again. The recombination reaction, like many others, speeds up as the temperature rises. Because the darkness of the glasses at any moment is a balance between UV light-induced disassociation and the temperature-sensitive reassociation, it takes much more UV to reach a given level of darkening in a warm climate.

ALEC CAWLEY

Photochromic materials are sensitive to temperature and darken more when they are colder. My sunglasses turn really dark on an overcast day but change little in the midday sun of Florida. This is fine for skiers but not much use to sun lovers.

I also found, to my cost, that many photochromic lenses react almost entirely to UV radiation rather than to visible light, so they don't darken properly inside a car.

WILLIAM DARLINGTON
BELL COLLEGE OF TECHNOLOGY,
HAMILTON, STRATHCLYDE, UK

The response of photochromic lenses to light is affected by temperature. Lower temperatures change the kinetics of the photochemical reaction, so the reverse reaction—lens lightening—is delayed.

Photochromic lenses become much darker at lower temperatures. Living in the American Midwest provides me with perfect experimental conditions to test the temperature effects. With summer temperatures around 86°F my photochromic lenses respond with a bluish-gray tint, whereas in deep winter, at around 14°F, they quickly become very dark.

The darker lens tint on sunny winter days is especially beneficial against strong snow-dazzle. However, this heavy darkening is disconcerting when I go indoors on a sunny day because it takes about 10 minutes for the lenses to return to normal.

BARRY TIMMS
UNIVERSITY OF SOUTH DAKOTA,
VERMILLION

8 Troublesome Transport

◈ Stop-Go

Why are the colored lights in traffic signals universally arranged red over amber over green, as opposed to the universal practice for railroad signals, which have green over amber over red (for a three-aspect signal)?

ROGER HENRY

The difference between road and rail usage derives from the history of railroads and the primacy of safety. The old mechanical railroad signaling arms were designed so that failure, which would be in the "down" position, meant stop. The illuminated part of the signal consisted of two colored glass panels in the far end of the signal arm, beyond the pivot, which moved in front of a fixed lantern. Even though the higher of the two glass panels was the red panel, it showed when the signal was down, and this meant stop. While railroads retained mixed mechanical and electrical signaling the signals had to be compatible. Therefore, the new electrical signals showed red at the bottom so that train operators always equated either signal in its down or bottom position with the order to stop.

Road signals had no mechanical forerunner and are designed so the most important light, the red, can be seen from the greatest distance. This means putting it as high as possible. Additionally, visibility for railroad signals is not the same issue as it is on the roads. Railroad signal sites are carefully selected.

GERALD DOREY

Gerald Dorey is only partly correct in his historical explanation for the order of railroad signal lights. Indeed, he overlooks the large parts of the country where lower-quadrant semaphore signals (in which horizontal means danger and 45° down means clear) were used. In these signals the red light was therefore at the top.

The main reason for putting the red light at the bottom in modern British signaling installations is the weather. To ensure visibility in bright sunshine, each color light has a long cowl or hood above it. In the winter, however, snow can build up on these cowls and obscure the light above. Being at the bottom, the most safety-critical red light has no other light below it and therefore no cowl, so snow cannot build up to cover a red light.

VINCENT LUTHART

There are two kinds of mechanical, or semaphore, signals. In the older lower-quadrant type, the arm slopes downward from the pivot to show clear or green and is returned to the horizontal by a counterweight, and the lamp glasses are red above green. In the newer upper-quadrant type, the arm slopes up for clear and returns by its own weight (as in the scene in the classic film *The Lady Killers*), and the lamp glasses are in fact side by side. Red is nearer the pivot and green is to its right on the outside.

In both, the horizontal arm means stop, but this is not synonymous with down, which means opposite things in the two cases. Red arms are always used in stop signals, but distants (meaning warning) operate similarly. However, on distants the arm glass and lamp glass are yellow, not red, and these mean "pass with caution."

The arrangement of multiple-aspect, colored light signals

has nothing to do with that of arm position. The red is at the bottom simply because it is the position nearest the driver's eyes; yellow is above, then green, and in four-aspect signals the second yellow is topmost, above the green.

C. C. THORNBURN
ASTON UNIVERSITY, BIRMINGHAM, UK

Road users do not have to pass a color vision test, and therefore the position of the red, amber, and green lights must always be the same, so the light illuminated can be recognized by position as well as color. Such signals are usually placed at sites where a speed limit applies and, because of the higher braking coefficient of rubber tires, the driver can still stop safely even after identifying a red indication only by its position.

The train driver, whose color vision is checked regularly, has to act on signals at a far greater distance to ensure that the train can be stopped in time. On main lines the indication has to be identified accurately at long range, where it is impossible for the driver to see its position and where he must rely solely on its color.

The original question was, in fact, incorrect, because there is no universal railroad layout of green over yellow over red (in railroad parlance, the caution indication is referred to as yellow, not amber). In the past, some signals had only a single lens, the different indications being given by interposing colored filters over the beam. The only fixed rule with the layout of signals with multiple lenses is that the one which exhibits the red indication is mounted nearest to the line of the driver's eyes. In some places, therefore, it may be at the top, as it is with a road traffic signal.

On high-speed lines, it is necessary to have a double-yellow indication, which is exhibited by the signal before the one showing a single yellow for caution. That, in turn, will be a further three-quarters of a mile or so before the one showing red for stop. This gives a run of two signals warning of a stop signal ahead. Such double-yellow signals normally have the two yellow lights separated by the green one, to maximize their visual separation when they are viewed from a distance.

P. W. B. SEMMENS

 # Pressure Situation

We are all familiar with the popping ears associated with take-off and landing in an airplane. This is caused by changes in pressure. Given that the aircraft cabin is artificially pressurized, why isn't the internal pressure maintained at one level throughout the journey?

CRAIG LINDSAY

For reasons of fuel economy, large commercial passenger aircraft have to fly at altitudes far in excess of those capable of sustaining life. Whereas 18,000 feet is about the maximum altitude at which a person can live for any extended period, a subsonic passenger jet has the best fuel economy when flying at around 39,000 feet.

Aircraft manufacturers, therefore, have no choice but to pressurize the interior of a passenger aircraft. This poses huge technical problems. At 39,000 feet, where the pressure is about one-fifth of that at sea level, the pressure inside is trying to burst the fuselage apart. This pressure has to be contained, and all the stretching and flexing of the fuselage during a flight has to be kept within safe limits. It is far easier to do this if the pressure differential between inside and outside is kept to a minimum, so that a cheaper and lighter fuselage structure can be used.

For commercial airliners this means that the pressure inside during cruising is kept at the lowest possible safe level— 8,000 feet. This is about the maximum altitude which a normal healthy person can be subjected to without ill effects. Even at this altitude, unfit people, those with respiratory illnesses, and those who have sampled a few too many duty-free drinks might still feel ill.

There is another problem: not all airfields are at the same altitude. In an extreme case, a flight from Heathrow in England to La Paz in Bolivia would entail going from sea level to around 17,000 feet, where the air pressure is about half that at sea level. Under these circumstances it is just not possible to maintain the same pressure throughout the flight. Imagine what would happen if the pressures inside and outside were not the same at the time the doors were opened: the effect would be quite spectacular and most undesirable.

As for the ear popping, nowadays, "for your safety and comfort," the internal pressure is imperceptibly reduced, all under computer control, as the aircraft climbs. It is gradually increased (or, in the case of La Paz and other high-altitude airports, decreased) during descent so that, as the aircraft is coming to a stop on the runway, the pressure inside and out is the same. This is normally sufficient for your ears to adjust, but if all else fails, pinch your nose and gently but firmly increase the pressure in the nasal cavity until you feel the pressure equalize.

TERENCE HOLLINGWORTH

An advantage of flying by Concorde was that the fuselage had to be especially strong to fly at very high altitudes, so the cabin pressure did not have to be reduced below that experienced at 3,000 feet.

ARTHUR COX

Porthole Paradox

Why are the windows of a ship's hull round? And when did this design begin?

CAMPBELL MUNRO

I assume that your correspondent is referring to old pictures and prints of wooden ships, where the portholes (usually gun ports) are square or rectangular, and is wondering why such ports are round in steel-hulled ships.

When ships were made of wood, the architectural material was fibrous and fairly flexible (wooden ships really did creak, and this creaking was caused by timber flexure from wave action). However, wood—especially wet wood—is highly resistant to fatigue stress. Try breaking a piece of wet willow by repeated reverse flexure, and then try the same process with a soft milled steel bar or rod of similar section. Ferrous metals (indeed, most metals), are highly prone to crystalline fracture as a result of changes to the grain structure arising from repeated stress reversal. The effect depends upon section, heat treatment, carbon content, and any alloying elements present.

Toward the end of the nineteenth century, steel hulls became universal for merchant vessels, and subsequently for warships. Naval architects found out pretty quickly that any rectangular or square hole in a ship, whether on a deck (a hatch) or in the hull (a porthole) was a source of metal fatigue, commencing at the corners. The hull or deck would literally rip, owing to flexure cycles brought about by wave action; the rougher the seas, the greater the magnitude of the stress. The unlucky sailors found that their ship was most likely to fall apart when weather conditions were at their worst. Thus, naval architects specified circular portholes, and rounded corners for deck hatches. This left no sharp corners for stress concentration.

DAVID LORD

 # Splat!

The following paradox has puzzled me since I was a child. A fly is flying in the opposite direction from a moving train. The fly hits the train head-on. As the fly strikes the front of

*the train, its direction of movement changes through 180°,
because it hits the windshield and continues as an amorphous blob of fly-goo on the front of the train.*

*At the instant it changes direction, the fly must be sta-
tionary and since, at that instant, it is also stuck onto the
front of the train, the train must also be stationary. Thus a
fly can stop a train. Where is the logical inconsistency in this
(or does it explain something about British railroads)?*

GEOFF FLEET

You are right. A fly does stop a train, but not the whole
train, just part of the small local area where the fly makes
contact, and then not for very long.

All objects, no matter how rigid they seem, are flexible to
some extent. So the train's windshield, on being struck by the fly,
deflects backward very slightly. That small piece of train not
only stops for a short period but can actually move backward.

It takes considerable force to do this (glass being fairly
rigid), but we should remember that the forces involved in
any type of impact are typically quite large.

The force exerted by the fly on the train is the same size
as the force exerted by the train on the fly—a large force.
And such a force acting on the small mass of the fly gives
rise to a very large rate of acceleration. In fact, the rate of
acceleration of the fly is so great that it accelerates up to the
speed of the train in only the short distance by which the
windshield has been deflected.

Having gotten the fly up to speed, the windshield then
springs back to its original shape. Because it springs back
very quickly, the deformed part actually overshoots its original position and a vibration is then set up as it springs back
and forth trying to regain its original form. This gives rise to
the sound we hear when the fly hits the windshield.

This simple picture is complicated by factors such as the
crushing of the fly's body and inertia effects in the glass, but
it does demonstrate the principles that are involved.

ERIC DAVIES

The questioner is correct in the assumption that the fly must, at some point, be stationary. However, at this point it is not "stuck" onto the front of the train.

As soon as the train's front window touches the front of the fly (if we ignore the effect of the wall of air that is pushed in front of the train), the fly is accelerated forward in relation to the train. During the very short, but finite, period of time that it takes the train to cover the length of the fly's body, the fly is being compressed and accelerated. Thus, in the instant that the fly is stationary, perhaps its front 10 percent has become goo on the train's window. The train has maintained a constant speed during this process. By the time the front of the train has completely caught up with the whole of the fly, some 2×10^{-4} seconds later at 130 miles per hour, the fly has been accelerated up to the speed of the train and continues, now completely flattened, to move with it.

A slightly more pedantic point is that, by conservation of momentum, the train will be very slightly slowed, although it will quickly build up to its original speed. The acceleration felt by the fly, if accelerated by 130 mph over a quarter-inch, is around 10^6 feet per second per second—about 30,000 g. The force felt by a fly and the window is around 300 newtons.

JULIAN BEAN

When the train hits the fly, the front few nanometers of the windshield's impacting surface stop momentarily, the next few nanometers suffer elastic deformation, and the rest of the train continues at full speed.

After the impact, the compressed windshield material will recover, accelerating its front edge up to the full speed again, and showing virtually no sign of damage (unlike the inelastic deformation of the fly).

This is a slight oversimplification, as in practice an elastic stress wave will propagate backward into the train, and the front surface will oscillate until the motion is canceled out, but such effects will be unimportant in the case of the fly and the train. Where the masses are more equal, as in the case of colliding cars, the additional motions within each structure may be important: for example, they may determine the type of injuries suffered by the occupants.

M. G. LANGDON

Readers' explanations about the fly hitting the train cover many aspects, from the width of the fly to the pliability of the windshield. (What if the fly hits the boiler instead?)

But they completely miss the implied point of the question, which is philosophical rather than physical. For "fly" substitute "one atom of the fly." This is just a rerun of the paradox posed by Zeno of Elea. Around 450 BC he said that a moving object is always in motion, and yet at any given time it is somewhere (that is, stationary). We humans cannot see, measure, or imagine an infinitely small time any more than we can truly imagine infinity. We never will.

R. K. HENDRA

✧ Through the Hole

I recently did a parachute jump for charity, and the one disconcerting thing about the jump (apart from my fear of heights) was the large hole at the top of the parachute. Why is it there? Does it help in any way to reduce the drag on the chute?

SUZY KLEIN

In the days before the apex vent (the disconcerting hole in the top of the parachute canopy), the only way that the air trapped underneath the parachute could escape was to spill out from one edge of the canopy, thereby tilting it and throwing the helpless parachutist to one side.

As the canopy swung back, more air would spill out from the opposite side, setting up a regular, pendulum-like oscillation (watch any footage of Second World War parachutists and you will see this).

As you can imagine, hitting the ground during a down-swing was hazardous, especially if it was also a windy day. The apex vent, by allowing the air to leak slowly out of the top of the parachute canopy, prevents this wild oscillation and makes for much safer landings.

Another benefit of the apex vent is that it slows down the opening of the parachute. Without the vent, air inflates the canopy much more abruptly, and it can damage the parachute or bring tears to the eyes of (particularly male) jumpers.

PAUL DEAR

Lean View

Why do airplanes have such small windows, and why are the windows positioned so low in the fuselage that most people have to bend down in order to see other airplanes on the runway?

TIMOTHY KOULOUMPAS

As with many things concerning the design of an aircraft, the final arrangement of various parts is based upon a series of compromises. An aircraft designer's life would be much easier if there were no windows at all, but so far the consensus seems to be that we should have them.

Britain lost the initiative in manufacturing jet airliners when the development of the de Havilland Comet in the 1950s suffered a setback after a series of crashes, in part because metal fatigue around the windows led to structural failure.

While windows remain an accepted part of aircraft design, they have since been kept as small as possible. These days they are typically 13 inches high. The window has to have three panes: two pressure panes and one interior pane to prevent passengers from getting at and damaging the vital ones. The panes are contained in a window unit, which is fastened and sealed to the aircraft structure.

It is, of course, much heavier and costlier than the thin sheet of aluminum it replaces, and the structure of the aircraft needs to be reinforced to support it. All this extra weight means that fewer passengers or less cargo can be carried, so it reduces airlines' potential revenues.

Windows also present a maintenance problem. As well as getting scratched and broken, they are a source of air leaks from the cabin and they also suffer from condensation and icing.

The position of the windows varies depending on the aircraft, but generally designers try to place them with their center line a little below the eye level of seated passengers. On the ground this is perhaps too low, but in flight it gives an oblique view of the ground. Little would be gained by positioning the window higher. Because the seats are placed at the widest part of the circular or oval fuselage, the windows would be angled upward some 10 or 15 degrees. The only view the passenger would then have in flight would be of the sky. Also, if the top of the window were above eye level there would be a constant problem from the sun's glare and dazzle. Passengers would just end up pulling down the blinds, negating the benefit of having a window in the first place.

It would be useful to have deeper windows; but, as I have already said, the weight penalty makes this impractical.

It also has to be remembered that every commercial aircraft flying today was designed at least 10 years ago, and some actually started life on the drawing board 40 years ago. During this time people have changed and seat design has changed. When these aircraft were developed, the structural design—including the position of the windows—was fixed, and the window line has traditionally been used as a convenient breaking point to bring pieces of fuselage shell together. This position having been determined, and production lines then set up with the correct tooling, it would be enormously costly to change it.

In the meantime, people have been getting bigger. Designers have to use what are known as "Dreyfuss criteria" to determine seat sizes. These criteria are continually changing, but a designer will typically make a plane's seats big enough to accommodate 95 percent of American males. If you are particularly tall, this is going to make the window seem lower for you—and people are generally taller than they used to be.

Finally, the present trend in air travel is away from luxurious, spacious layouts to high-density seating. In these circumstances, where the seat pitch is reduced to accommodate as many passengers as possible, the seat base has to be higher to provide legroom for the person sitting behind. This also makes the relative window position lower still than was originally intended.

TERENCE HOLLINGWORTH

The windows on aircraft are so small to make them safe. The first major jet airliner, the de Havilland Comet, had large, rectangular picture windows through which the passengers had a great all-around view. But after a few years in service, the aircraft started to break up during flight.

To find out why, de Havilland put a new Comet into a tank of water and then pressurized and depressurized it repeatedly to simulate the conditions of flight. After the equivalent of two years' worth of pressurization cycles (which actually took only a few weeks in the water tank), the airframe was found to fail in the top corner of one of the large windows, causing a catastrophic breakup in flight.

The windows had to be redesigned, and small round windows set low in the fuselage were created. This solved the problem, and the position of the windows remains the same today.

MIKE BURNS
WELLINGTON COLLEGE
CROWTHORNE, BERKSHIRE, UK

⬡ On the Turn

Why, when you are driving, does the steering wheel of a car straighten itself if you remove your hands after turning it? It doesn't happen on my friend's Lego Technics car.

CLARE SUDBERY

The tendency of the steering wheel to straighten itself is caused by the caster action of the front wheels. This effect is more clearly seen on a shopping cart, where the vertical swivel axis of each wheel is in front of the wheel-to-ground contact point. If you start pushing the cart when the wheels are not aligned to the direction of the cart's motion, the wheels are pulled around into alignment by the drag force between ground and wheel.

The full explanation is that as the cart moves forward, the drag force exerted by the ground on the wheel always opposes any relative motion (or slip) between the wheel and the ground.

Unless the wheels are aligned to the cart's motion, the drag force does not pass through the swivel axis, and therefore it produces a turning moment around that axis which always acts to bring the wheel back into alignment.

In a car, the same effect is achieved by inclining the steering axis and ensuring that the point where the axis intersects the ground is ahead of the tire–ground contact point. The same is true of a bicycle, as you can see if you hold a broom handle alongside the steering axis of the bike so that the handle touches the ground. You should see that this point is just in front of the tire–ground contact point.

You can demonstrate the caster action on a bicycle for yourself by pushing the bike backward and forward by the saddle while the handlebars are left free. When you are going forward, the bike is easy to push in a relatively straight line.

However, going backward is almost impossible because

the front wheel tries to turn around through 180 degrees just like a shopping cart wheel. You will also find when reversing a car that the steering wheel loses its tendency to center itself.

BILL LAUGHTON

Which Way Is Up?

My whole class, including my mathematics teacher, is baffled. We cannot work out how an aircraft can manage to fly upside down without crashing into the ground. We understand that the wings are designed to provide uplift when the plane is flying horizontally. However, when the plane flies on its back as some smaller jets often do, surely the uplift is working in reverse and forcing the plane back down toward the ground. Yet most types of small aircraft seem able to maintain the upside-down position for long periods of flight. How do they do this?

NIK YUSOKK

Although the airfoil shape of an aircraft's wing produces some of the lift in normal flight, the more important factor is the angle of attack—the angle at which the air strikes the wing.

The wings of an aircraft are normally inclined to about 4° to the horizontal when compared with the main body of the aircraft. This is known as the chord angle of the wing.

So even when the fuselage is level, the angle of attack into the oncoming wind is 4°. This produces lift in the same way that your hand experiences an upward force when you hold it at about 45° to the horizontal in a fast-moving stream of air. Your hand does not have an airfoil shape, but the lift that you feel is caused by the angle of attack of your palm to the oncoming wind.

It is this principle that allows an aircraft to fly upside

down. The nose is pointed farther upward than in standard flight because of the need to offset the chord angle of the wing. But if the angle of attack is positive compared with the relative airflow over the wing, then an upward force will still be produced. It is this lifting force that overcomes the force produced by the shape of the wing, and holds the aircraft in the air.

The bigger problem that pilots should be concerned about when flying their aircraft upside down is the risk of the engine's stopping, because the oil and fuel systems in most ordinary light aircraft are fed only by gravity. Flying your aircraft upside down can easily cut off the fuel supply because the valve that is feeding fuel to the engine suddenly finds itself at the top of the tank.

MARK MOBLEY

Mercury Rising

On a recent flight, I was studying a card listing items that were prohibited by airlines. I was amazed to see that I couldn't take a mercury thermometer on a flight. Why on earth not?

RICK ERAHO

Planes are largely made from aluminum and, surprisingly, a very small amount of mercury can destroy a large amount of aluminum. Despite its apparently inert behavior, aluminum is actually a rather reactive metal that will combine violently with oxygen in air. However, this reaction quickly produces a thin, tough oxide layer, which stops further attack. The process of anodizing the aluminum thickens this layer to give better protection.

Mercury has the ability to disrupt this protective oxide layer, and the results can be spectacular. It can dissolve aluminum to form an amalgam that may break up the oxide layer from below—presumably the initial attack occurs through tiny faults in the oxide.

Many years ago a technician working for me spilled a few drops of mercury on his wooden bench, which had heavy aluminum angles screwed around the edges to protect it. Next morning large holes were eaten through the aluminum, the wood nearby was deeply charred, and large fragile towers of friable aluminum oxide had grown like strange corals.

This used to provide a fine chemistry experiment, but it is now frowned upon because of the toxicity of the mercury.

On one occasion a passenger in front of me was prevented from carrying a barometer onto an aircraft because it was on the list of prohibited articles, even though this particular barometer was empty. With difficulty I persuaded the staff that it was harmless. They did not realize it was the mercury that was dangerous; they thought it was just barometers per se. I wonder what they thought an altimeter measures . . .

HARVEY RUTT
DEPARTMENT OF ELECTRONICS AND COMPUTER SCIENCE
UNIVERSITY OF SOUTHAMPTON, UK

Given the mobility of liquid mercury, the corrosive amalgam may form deep within the structure. An aircraft in which mercury has been spilled must be put into quarantine until the amalgam makes its presence known. Ultimately, the aircraft is likely to be scrapped because the engineering textbooks state that the amalgam spreads slowly like wood rot to adjacent areas.

ROD PARIS
AIR MEDICAL LIMITED
OXFORD AIRPORT
KIDLINGTON, OXFORDSHIRE, UK

Mercury, along with many other common chemicals, is classified under "dangerous goods" in international regulations developed by the International Civil Aviation Organization (ICAO), which is part of the UN. You are not permitted to carry this substance, or any article containing it, aboard an aircraft in hand luggage or checked-in baggage. An exception is made for small clinical thermometers in protective cases for personal use.

Should mercury-containing articles need to be transported, they must be consigned as air freight. The ICAO rules specify in detail how this must be done.

Don't think that you can afford to ignore these restrictions. In Britain, endangering an aircraft by taking aboard dangerous goods could result in a charge and a hefty fine under the 1982 Civil Aviation Act. In the event of a mercury spillage the aircraft would need to be taken out of service. The airline and/or its manufacturer may try to recover costs from you or your employer.

James Hookham
Freight Transport Association
Tunbridge Wells, Kent, UK

◈ Escalating Concerns

I have noticed that when I travel on an escalator, the handrail always moves at a different speed from the stairs. You would expect it to move at the same speed, but it never does. Why not?

Bernd Haupt

The stairs and handrail are designed to move at the same speed and are driven by the same electric motor. The motor connects to a drive gear that moves the steps, and from there a belt turns a wheel that drives the handrail. Although the handrail will ideally move at the same speed as the steps when first installed, it wears and stretches as it is used. As a result, it can change speed. Improper handrail setup, seized rollers, flat spots, and contaminants on the handrail drive surfaces can all affect the speed.

JOHAN UYS

The handrail can travel at different speeds but it is not supposed to. The American National Standards Institute code ANSI A17.1 requires that the speed of the handrail shall not change when 444.8 newtons are applied against the direction of motion. To meet this requirement, the handrails are sometimes adjusted to move slightly faster than the step. Escalators that are installed under the ANSI A17.1-1990 code require that a handrail-speed monitoring device be installed. If the speed of the handrail changes by more than 15 percent, all power is removed from the motor drive and the brakes are applied.

RICHARD A. KENNEDY
RICHARD A. KENNEDY & ASSOCIATES
MAINTENANCE AUDITORS FOR ELEVATORS AND OTHER LIFTING DEVICES
WEST CHESTER, PENNSYLVANIA

Escalator handrails are moved by the friction of a rubber-tired wheel operating against the inside of the handrail, and slippage is not uncommon, though seldom uniform. Most commonly a buildup of oil and dirt on the inside canvas of the handrail causes some slippage, although this can be cleaned off and the canvas roughed up to afford more traction. Passengers pulling on the handrail will also make it slip.

The handrail drive runs off the step drive, so the handrail should always match its speed. The typical diameter of a handrail drive wheel is between 3 and 4 feet, so one-tenth of an inch of wear on the drive tire would result in the loss of roughly one-twentieth of an inch of handrail travel per foot of step travel, which is hardly noticeable.

Other possible causes of handrail slippage include bald patches on the drive tire or the handrail, either of which would make the slippage highly predictable. In rare cases, with particular manufacturers, the drive chain can stretch to such a degree that it is forced to jump over a cog or two. This results in a loud noise and a noticeable jerk in the handrail.

GEOFFREY WOOD

British Standard EN115: 1995 states that the handrail speed should match the step speed to within 2 percent. The system driving both the step and the handrail are derived from the same source, so in theory they should run at the same speed. In practice, the step system consisting of precision-made metal components, allows the step speed to be easily and accurately controlled. In contrast, power is transferred to the handrail by friction, and the rubber and neoprene components make the system vulnerable to slippage and stretch from loading and frictional losses. These factors make it more difficult to accurately control the handrail speed—hence the necessity in British Standards to allow a 2 percent tolerance.

In reality, a small degree of slippage actually increases the level of safety should anything obstruct the handrail system.

BHAMINI GORE
OTIS LIMITED
LONDON, UK

⚙ Water Wheels

When you are driving a speedboat, why don't you have to change gears when you change speed, as you do in a car?

GRAHAM LUNDEGAARD

Some powerboats do have gears, but these are the exception rather than the rule—Ed.

The difference between boats and cars lies in the way the power generated by the engine is translated into movement of the vehicle. In a boat the engine turns a propeller, which pushes water backward. The reaction to this rapidly moving stream of water pushes the boat forward.

If the engine and propeller are well matched, there will be sufficient power to turn the propeller, even when the engine is running slowly. If the boat is large it may take some time to accelerate, during which water can be seen streaming away from the stern. Have a look next time you are on a ferry; you'll see churning water at the back of the ship, even though the ship has yet to move.

In a car—in contrast to the ferry—the wheels can turn only if the car moves but it takes a lot of power to accelerate from standing. Unfortunately, internal combustion engines do not generate much power when they are running slowly, so if the engine were connected to the wheels without a gearbox, the inertia of the vehicle would stall the engine. The gearbox allows the engine to turn rapidly, generating power, even when the wheels are moving slowly. If it were not for the ingenuity of gearbox and clutch designers, the internal combustion engine would have had no future in road vehicles. In contrast, steam engines generate a lot of power from a standing start, so steam locomotives can pull away without a gearbox.

On loose surfaces such as sand, a car's wheels can turn without the vehicle immediately moving, too. This situation is somewhat similar to the ferry, in that sand is thrown backward as the wheels rotate. However, more sand does not immediately rush in to take its place, so the wheels are likely to dig themselves into the sand until the car is embedded to its axles.

JOHN GEE

Speedboats experience a huge amount of drag. Typical full-speed drag force is a quarter of the weight of the boat, which is like driving a car up a slope of slightly over 25 percent. Speedboats have to be low-geared to overcome this drag, and a multispeed gearbox would make very little difference for low-speed acceleration. The drag is so large that any gear change would have to be extremely fast or the boat would slow down too much during the changes. Because propellers slip through the water when the boat is starting up, there is no need for a clutch—the water acts like one.

It is possible to change gears on a boat by changing the propeller for one of a different pitch. A lower pitch gives better acceleration, and allows you to pull larger loads, while a higher pitch gives better top speed, if the waves are small, which reduces drag. However, the typical change that might be useful on a boat is less than the difference between adjacent gears on a car.

MALIN DIXON

The speed of a car is proportional to the engine speed for a particular gear. This is not the case in a boat, because the propeller can "slip" in the water, whereas a car tire stays stuck to the road. In all engines an increase in revolutions means an increase in power, up to a certain point.

Most of us will have accidentally taken off from traffic lights in a car in third gear. The number of revolutions in third gear is much lower than that in first gear, so the engine does not generate enough power to move the car, and it stalls.

This shows that a low gear is essential in a car for power at low speed. But if you open the throttle fully in a boat, the propeller spins freely in the water, the engine reaches a high number of revolutions, and the boat moves off without stalling. The boat's single gear is designed so that the propeller works most efficiently within the engine's operating range. There is no need for additional gears.

In a boat, the drop in power while you are changing gears results in a large drop in speed, because the resistance in water is much greater than on a road, so a boat cannot pass through a gear change as easily as a car can.

DAVID EDELMAN

9   Best of the Rest

Gung-Ho Guns

In many parts of the world, people celebrate victories, birthdays, and similar events by firing guns into the air with great exuberance and a seeming disregard for the welfare of themselves and others. Assuming the barrel of the gun is perpendicular to the ground when the bullet leaves it, approximately what altitude would it reach and what is its velocity (and potential lethality) when it falls back to Earth?

LEO KELLY

Firing handguns into the air is commonplace in some parts of the world and causes injuries with a disproportionate number of fatalities. For a typical modern 7.62 millimeter-caliber bullet fired vertically into the air from a rifle, the bullet will have a velocity of about 2,750 feet per second as it leaves the muzzle of the gun and will reach a height of about 7,800 feet in some 17 seconds. It will then take another 40 seconds or so to return to the ground, usually at a relatively low speed, which approximates the terminal velocity. This part of the bullet's trajectory will normally be flown base first, since the bullet is actually more stable in rearward than in forward flight.

Even with a truly vertical launch, the bullet can move some distance sideways. It will spend about 8 seconds at between 7,400 and 7,800 feet and at a vertical velocity of less than 130 feet per second. In this time it is particularly susceptible to lateral movement by the wind. It will return to the ground at a speed of some 230 feet per second.

This sounds quite low; but, because of the predominance of cranial injuries, the number of deaths and serious injuries as a proportion of the number of gunshot wounds is surprisingly high. It is typically some five times more than is observed in normal firing. As might be expected, measurements of this kind are rather difficult, and the above values come from a computer model of the bullet flight.

SAM ELLIS AND GERRY MOSS
ROYAL MILITARY COLLEGE OF SCIENCE
SWINDON, WILTSHIRE, UK

Different bullet types behave in different ways. A .22 long-range bullet reaches a maximum altitude of 3,868 feet and a terminal velocity of either 197 feet per second or 141 feet per second depending upon whether the bullet falls base first or tumbles.

A .44 magnum bullet will reach an altitude of 4,517 feet and will have a terminal velocity of 249 feet per second falling base first. A .30-06 bullet will reach an altitude of 10,105 feet with a terminal velocity of 324 feet per second.

The total flight time for the .22LR is between 30 and 36 seconds; for the .30-06 it is about 58 seconds. The velocities of the different bullets as they leave the rifle muzzle are much higher than their falling velocities. A .22LR has a muzzle velocity of 1,256 feet per second and the .30-06 has a muzzle velocity of 2,700 feet per second.

According to tests undertaken by Browning at the beginning of the twentieth century and recently by L. C. Haag, the bullet velocity required for skin penetration is between 150 and 200 feet per second, which is within the velocity range of falling bullets. Of course, skin penetration is not required in order to cause serious or fatal injury, and any responsible person will never fire bullets into the air in this manner.

The questioner may like to read "Falling Bullets: Terminal Velocities and Penetration Studies," by L. C. Haag, Wound Ballistics Conference, April 1994, Sacramento, California.
DAVID MADDISON

John W. Hicks, in his book *The Theory of the Rifle and Rifle Shooting*, describes experiments made in 1909 by a Major Hardcastle, who fired .303 rifle rounds vertically into the air on the River Stour at Manningtree. His boatman, probably a theorist unaware of the winds aloft, insisted on wearing a copy of *Kelly's Directory* on his head.

However, none of the bullets landed within 100 yards; some landed up to a quarter of a mile away and others were lost altogether.

Julian S. Hatcher records a similar experiment in Florida immediately after the First World War. A .30-caliber machine gun was set up on a 10-foot-square stage in a sea inlet where the water was very calm, so that the returning bullets could be seen to splash down. A sheet of armor above the stage protected the experimenters. The gun was then adjusted to center the groups of returning bullets onto the stage.

Of more than 500 bullets fired into the air, only four hit the stage at the end of their return journey. The bullets fired in each burst fell in groups about 25 yards across.

The bullets rose to approximately 9,000 feet before falling back. With a total flight time of about a minute, the wind exerts a noticeable effect on the return point.
DICK FILLERY

In my youth, I used to collect, for salvage, brass cartridge cases ejected from aircraft machine guns during the Battle of Britain. They drifted down slowly from the sky because, I guess, their ratio of mass to surface area was low. However, they were still warm when I picked them up.

Accordingly, if the projectile is small, like a .303 bullet, it does nobody much harm when it lands. Like that of a mouse in a mine shaft, its terminal velocity is negligible. However, if

because of its mass the projectile has enough terminal velocity, it could kill you.

M. W. EVANS

Aisle Miles

*Two people lose each other while wandering through the
aisles of a large supermarket. The height of the shelves pre-
cludes aisle-to-aisle visibility. One person wishes to find the
other. Should that person stop moving and remain in a single
visible site while the other person continues to move through
the aisles? Or would an encounter or sighting occur sooner
if both were moving through the aisles?*

DAVID KAFKEWITZ

The best strategy may be to wait at the exit of the store, on
the assumption that the other person may eventually conclude that you have gone home and do likewise. The maximum waiting time will then be from the time you lost each
other until the store closes.

A strategy of staying still works only if just one person stays
still. If you both decide to stay still, then the wait time is either
infinite, if you get locked in, or again until the store closes.

Assuming that one person stays still while the other
searches, then the maximum time is the time taken for one
person to search the entire store. This depends on the layout
of the store: if all the aisles can be readily seen from one van-
tage point, then the search is simplified. The problem is not
unlike that of designing prisons in which the guards can see
down as many corridors as possible, or the design of forts
that will give the defenders maximum cover. In order to increase the odds of being located, the person staying still
should stand at an intersection of aisles.

A random search will proceed with each person moving
away from his or her initial starting point at a rate propor-

tional to the square root of time. The area being searched by each person is defined by two circles centered at their respective search starting points. Given that these circles will need to overlap significantly for the individuals to meet, the search time must be at least proportional to the square of their initial separation distance. If some of the aisles are blocked during this search, then the rate of movement is reduced and the problem becomes one of motion on a fractal where the rate is proportional to some fractional exponent.

STEPHEN MASSEY

To begin to answer this question, one must first know whether the two people have agreed in advance what to do if separated—for example, who should wait and who should search. If they can agree on independent search strategies in advance, the problem is the asymmetric version of the rendezvous search problem (see below); otherwise, it is the symmetric version.

I discuss both versions of the problem in a paper to be published in the *Society for Industrial and Applied Mathematics Journal of Control and Optimization,* and several specific cases in particular search regions have subsequently been solved. In all these cases where exact solutions (to give at least expected time or minimax time) have been obtained, both searchers move at their maximum speed all the time. In these cases it is certainly not optimal for a searcher to stop while the other continues. For example, in a simplified model in which two people are placed a unit distance apart, but neither knows the direction of the other, it would take an expected time of $(1 + 3)/2 = 2$ for the searcher to find the stationary person (assuming visibility is nil). However, by moving optimally the searchers can reduce this time to 13/8.

The only case I know of where a searcher and a waiter may be optimal is for two people placed randomly on a circle, and then only when the people concerned have no common notion of clockwise; otherwise, one person walking clockwise and the other counterclockwise is optimal.

All these results and questions assume that the searchers find each other only when they meet or alternatively when they come within a specified detection radius. This applies to aisles in a supermarket on a crowded day, when visibility along an aisle is limited. The possibility of seeing a long distance along an aisle has not, as far as I know, been modeled.

In case anyone is interested, the full bibliography on this topic is: "The Rendezvous Search Problem," S. Alpern; "Rendezvous Search on the Line with Distinguishable Players," S. Alpern and S. Gal; "Rendezvous Search on the Line with Indistinguishable Players," E. Anderson and E. Essegaier. All three papers appeared in the *SIAM Journal of Control and Optimization* in 1995.

STEVE ALPERN
LONDON SCHOOL OF ECONOMICS, UK

I recommend that you walk along the edge of the supermarket where the cash registers are, looking down the aisles for the person you seek. If you have no success, then walk back, still looking down the aisles, but also checking the cash registers. If you still have no success, then find the cold meat counter, as lines often develop there. Then have a final walk along the cash-register edge, checking the aisles again. If you are still unsuccessful, then you should ask for an announcement to be made on the public address system—or, if it's not urgent, wait by the exit.

OWEN CROSSBY

⚙ F-Factor

The following recently made its way into the New Scientist *office . . .*

> *Please read this sentence and count the F's:*
> *FINISHED FILES ARE THE RE-*

*SULT OF YEARS OF SCIENTIF-
IC STUDY COMBINED WITH
THE EXPERIENCE OF YEARS.*

*How many did you see? On first reading most people see
only three. However, the answer is actually six. Why is this?*
RICK ERAHO

The fact that most people can only see three F's instead of
six would be strange only if reading were entirely phonetic.
In reality, several methods are used to get meaning from
print, and the most common of these has little to do with individual letters and their sounds.

A reader becomes familiar with the shape of many words,
particularly short, common words like "of." These shapes
are memorized and the reader no longer sees the words as
made up of separate sounds. So, when you read the sentence,
you spot the F's in the longer, less familiar words but not in
the three occurrences of "of."
SAM HILL

An intelligent 7-year-old and a proofreader would read six
F's, because they have learned to give all words equal values.

As we learn to read faster, we select the most important
words, permitting our brains to fill in the gaps. The faster
we wish to read, the more words we must skip. It is quite
possible to read simple narrative at faster than 600 words
per minute, with comprehension. Fact-packed scientific text
obviously has to be read more slowly.

A fast reader must concentrate on the most important
words, usually nouns and verbs. Adjectives and adverbs
come next, with modals, articles, pronouns, and prepositions
and so on coming last. Experienced readers take less notice
of the least important words, which in English (as in most
other languages I know) tend to be short words. So all the
little words like "of" are skipped, and their F's are omitted
from the count.

I can only speculate abut reading speeds in Chinese or Japanese, where writing systems are quite different.

VALERIE MOYSES

I tried this on a colleague at work, without using the same layout of words and characters. When I rewrote it two occurrences of "of" were at the end of lines, and my friend spotted these F's but missed the remaining one. So I would assume that the way they are placed (in the middle of the line) has an effect on their noticeability.

Obviously, someone who does not understand English would have no problem spotting the F's. It would be interesting to see how someone who has English as a second language responds to this.

ANDREW MACCORMACK

I found out by reading an optical illusions book—*Can You Believe Your Eyes?* by J. Richard Block and Harold E. Yuker—that most people think there are three F's because they do not notice the F's in "of." This is because the F in "of" is pronounced as a V, so the brain doesn't recognize it as an F.

BRYN HART (AGE 10)

Great shame was felt, not only at seeing only three F's but at still seeing three after reading the answer. Then my wife sat down and announced that she saw six on the first reading.

I am an English teacher and my wife is a math teacher, and there lies the difference. The reader tries to understand the passage and casually ignores the word "of," which appears three times. The mathematical mind does exactly what the question asks—it counts the F's. My wife sees a number of letters but I see a phrase and therefore ignore the actual task.

The splitting of words with hyphens is also important because it puts pressure on the reader to understand the text and to follow the words, diverting attention from the real task.

TOM SWEETMAN

Your correspondent will be pleased to know that English is my second language (Taiwanese is my first) and I am due to take math A-level exams this summer. My F count on first reading was three, and it was still three after seeing the answer. I realized that the additional F's were in "of" only after I spelled it all out aloud. In fact, I realized this only as I recited the second "O. F."

It is my belief that your background has no bearing on whether you spot the F's or not. My training as a systems analyst should help me to concentrate on the task at hand (counting the F's). As to why I missed them, two explanations remain. Either it was too late at night or I am obviously not suited to doing a math degree and should contemplate a change of course.

ALEX LU

As a postscript I submit the following:

THE
SILLIEST
MISTAKE IN
IN THE WORLD

When trying to persuade a class of 7-year-olds of the need to reread what they had written, I put the above words on the blackboard. My best readers immediately read it as "The silliest mistake in the world" and I replied, "You've just made it." It was only when we reached the slower readers that the extra "IN" was discovered. At this point the principal came into the room, glanced at the board, and read the message aloud, but incorrectly. A chorus from the class greeted him: "You've just made it, sir."

The F and V confusion that was suggested by one of your correspondents cannot be the explanation here.

DOUGLAS BOOTE

☼ Dirty Business

In James Bond films, a gun with a silencer is used to dispose of bad (and good) guys. How does the silencer work?

JEREMY CHARLES

Silencers are more properly called sound moderators or suppressors and are widely used by hunters to reduce noise levels from the discharge of firearms, particularly sporting rifles and air weapons. A sound moderator is essentially no more than a series of baffles coupled to an expansion chamber, contained within a tubular attachment that screws onto the end of the firearm's barrel.

The noise of the discharge of most firearms is made up of two components. The first comes from the rapid expansion of propellant gases as they leave the muzzle. The second is the supersonic crack of the bullet. It is not possible to reduce the sound level of a supersonic bullet, but a sound moderator fitted to such a rifle will have some significant effect in reducing the noise signature because it controls the rate of expansion of the propelling gasses.

For a sound moderator to be really effective, it must be used with ammunition whose projectiles travel at less than the speed of sound. In such cases, the noise of the discharge is greatly reduced and may not even be recognizable as a gun.

It is not possible to fit a sound moderator to a revolver, because the gap between the barrel and the front of the cylinder means that about 5 percent of the propellant gases escape, contributing to the overall noise of the discharge. Otherwise, sound moderators can be fitted to any type of firearm.

I once watched a Second World War Sten submachine gun fitted with a large integral moderator being fired using special subsonic ammunition. The results were impressive: the only noise that came from the weapon was the clatter of its bolt.

In the public imagination, sound moderators for firearms invariably have a James Bond or underworld image. In reality, they are widely used in the countryside by hunters who wish to play their part in reducing noise pollution.

BILL HARRIMAN

BRITISH ASSOCIATION FOR SHOOTING AND CONSERVATION
WREXHAM, CLWYD, UK

The first successful silencers were patented in 1910 by the American inventor Hiram P. Maxim (son of Hiram S. Maxim of Maxim machine-gun fame). His devices were of the baffle type, which is still in common use today. A baffle silencer typically consists of a metal cylinder, usually divided into two sections, which is fixed to the muzzle of the firearm.

The first section, which is typically about a third of the silencer's length, contains an "expansion chamber" into which the hot gases that follow the bullet out of the muzzle can expand to dissipate some of their energy. The expansion chamber may contain a wire mesh cylinder, whose function is to break up the column of gas and to cool it by acting as a heat sink.

The second section consists of a series of metal baffles, with a central hole to allow the passage of the bullet. The function of the baffles is to progressively deflect and slow the flow of gas emerging from the expansion chamber, so that by the time the gases emerge from the silencer, their flow is cooler, they travel at low velocity, and they are silenced. A motorbike silencer works on exactly the same principle.

There are also variations on this theme: some designs consist entirely of baffles, whereas others are based entirely on one large expansion chamber. In fact, a plastic soft drink bottle can be made into a fairly efficient silencer that will work for a limited number of shots before it breaks up.

Silencers usually work best with cartridges that fire subsonic ammunition, since this eliminates the sonic crack which is produced by a bullet that goes faster than the speed of sound.

Some silencer designs slow the bullet to subsonic speed by means of ports cut into the barrel, with the ported section extending to protrude into the expansion chamber. These ports bleed off gas from behind the bullet, thereby reducing bore pressure and, eventually, the velocity of the bullet. In other designs, the baffles are made from an elastic material with a central hole smaller than the bullet. These "wipes" are pushed open by the passage of the bullet and close when it is past. The idea is that they further slow the exit of gas. Not surprisingly, the wipes can wear out rather quickly and can affect the accuracy of the bullet.

A second, but less common, type of silencer is the "wire mesh" design. These silencers usually have the same expansion chamber as the baffle type, but the baffles are replaced by a column of knitted wire mesh with a central hole for the bullet. Here, the wire mesh acts to disrupt the column of gas as in the baffle design, while at the same time acting as a heat sink to cool the hot gas and hence quiet it. Criminals have been known to improvise this type of silencer, using wire wool or steel pan scourers to form the mesh.

The very latest innovation in muzzle-mounted silencers is the so-called "wet" silencer (or "wet can" in the United States). These designs allow the use of water or a lubricating oil. On firing, the hot expanding gases are cooled, and therefore quieted, by the exchange of heat into liquid. Wet silencers allow the designer to produce much smaller or quieter designs.

An alternative approach to silencer design, which dispenses entirely with the muzzle-mounted silencer, has appeared from Russia. Instead it uses a special cartridge in which the bullet is pushed out by a propellant-driven piston. The piston is stopped by the neck of the cartridge, trapping the hot, noisy gas entirely within the chamber of the firearm.

It is fair to say that Hollywood takes great artistic liberties with silencers. Most real designs are very much larger than the cigar-tube-sized ones typically shown in films, and are usually much less simple to fit and remove. Despite what is shown in films, it is usually impossible to silence a revolver because the gap between the cylinder and the barrel allows gas to escape.

Finally, forget the distinctive "phut" produced by James Bond's silencer. Real designs are more likely to produce a muffled crack, or to sound like a car door being slammed.

HUGH BELLARS

⚙ Plastic Solution

Checkout operators the world over vigorously rub any malfunctioning credit and debit cards on the nearest available article of clothing. Does this actually serve any useful purpose?

PHILLIP CLEAVER

From my experience, a credit or debit card will fail to "swipe" correctly for one of three reasons.

First, something has permanently interfered with the magnetic strip on the card, so that the computer cannot read it. The cashier will have to type in the number manually, and a new card will probably need to be issued. Second, the machine is faulty and is unable to read the card.

However, the third reason the card cannot be read is the most common cause. Dust or dirt of some sort has collected on the magnetic strip. This obscures the information from the electronic reader. A quick wipe on your sleeve is all that is required to resolve this and, in the vast majority of cases, the card will swipe successfully at the second attempt.

There is no great mystery and no big science behind this practice, at least not that I am aware of. If you keep your cards in the card compartment of a purse or wallet, they should remain reasonably clean, and swipe easily on the first attempt. This should also eliminate the first problem, because they will be protected from anything that is likely to irreversibly damage the strip.

CHARLOTTE DADSWELL

There is one drawback to rubbing the magnetic strip, and it was something I often experienced as a supermarket supervisor. Rubbing the card can sometimes make it more difficult to read because it becomes charged with static electricity that can interfere with the electronic reader.

The instinct is to rub the card in an attempt to remove any dust that may have stuck to it may work in the short term, but the extra static charge the rubbing has generated will ensure that even more dust will cling to the card later on.

CISSY AZAR

Happy Returns

Why do boomerangs come back?

ADAM LONGLEY

A boomerang is like two spinning airplane wings joined in the middle. It is held almost vertically before it is thrown end over end. Because it spins in this way, the top wing actually goes away from you faster than the bottom wing. This makes the sideways push on the top wing (similar to lift on an airplane wing) stronger than that on the bottom wing, so the boomerang gets tilted over, just as you would be if someone pushed on your shoulder, and its flight pattern begins to curve.

Similarly, if you ride a bicycle and lean over, the bicycle will turn, eventually going in a circle. The boomerang does too.

ALAN CHESTER

Returning boomerangs work by a combination of aerodynamic and gyroscopic effects. A boomerang is essentially a rotating wing with two or more airfoil-shaped blades. It is thrown with its plane of rotation at about 20° to the vertical and so that it spins rapidly (typically about 10 revolutions per second), with the uppermost blades traveling in the direction of overall motion. Therefore, the blade at the top moves through the air faster than the lower one. The faster-moving blades generate more lift than the slower-moving ones. This produces an overall force in the direction of turn, plus an overturning torque.

The rotation of the boomerang makes it behave like a gyroscope. When the overturning torque occurs, the gyroscopic effect makes the boomerang turn (or precess) around a different, near-vertical, axis. This continuously changes the boomerang's plane of rotation, causing it to travel around an arc back to the thrower.

Other effects are also evident in the boomerang's motion, such as its tendency to lie flat as it returns to the thrower—its plane changes from 20° from the vertical initially, to horizontal at return. This is caused by a number of aerodynamic effects combined again with gyroscopic precession. The most significant effect is that the blades on the leading side of the rotating boomerang generate more lift force than the blades on the trailing side, because of the disturbed airflow on the trailing side. This again causes rotation, which leads the boomerang to spin toward the horizontal plane. An article by Felix Hess in the November 1968 edition of *Scientific American* explains this process in detail.

RICHARD KELSO AND PHILIP CUTLER
UNIVERSITY OF ADELAIDE, SOUTH AUSTRALIA

The simple answer to this question is that most boomerangs don't come back and were never intended to do so. The Australian Aboriginal people made the boomerang for hunting and fighting rather than for sport or play, so they did not make the so-called returning boomerang throughout most of the Australian continent. For them, the real returns of boomerang throwing came in the form of fresh food or the beating of an enemy.

I have seen the Warlpiri people throw a karli boomerang and hit a target at well over 100 yards. Particularly skilled users of the karli throw this deadly weapon with surprising ease. The Warlpiri also manufacture the wirlki (also known as the "hooked" or "Number 7" boomerang), which is used for fighting.

Across Australia, even in those areas where the boomerang is not made, there is nearly universal use of paired boomerangs as rhythm instruments in ceremonial contexts. Such boomerangs are still traded for ritual use across thousands of miles.

There are and have been an astonishing variety of boomerangs from Australia. For an accessible account see *Boomerang: Behind an Australian Icon* by Philip Jones, published by the South Australian Museum.

CHIPS MACKINOLTY

It's a Cracker

Why does the end of a whip crack?

DAVID INNES

The crack is actually a sonic boom, caused when the end of the whip breaks the sound barrier. This is possible because a whip tapers from handle to tip. When the whip is used, the energy imparted to the handle sends a wave down the length of the whip. As this wave travels down the tapering whip it acts on a progressively smaller cross section and a progressively smaller mass.

The energy of this wave is a function of mass and velocity, and since the energy of the wave must be conserved, it follows that if mass is decreasing as the wave travels down the whip, then velocity must increase. Therefore, the wave travels faster and faster, until by the time it reaches the tip it has attained the speed of sound.

MIKE CAPP

When the wave reaches the top of the whip it must be dissipated. Some goes to the air and some into a reflected wave that travels back up the whip. At the point that the initial wave reaches the tip and is about to embark on its return it undergoes a brief but enormous acceleration. The result is that it moves supersonically.

ANDREW PLANT

✹ Shedding Light

During a physics practical lesson, my tutor placed a lit candle on a turntable. When the table revolved we expected to see the flame on the candle point outward but, instead, it pointed inward. The school head of science couldn't explain this. Can anyone else?

RUTH HAVELAND

Yes, people can, but, despite a large number of replies, we found we had to knit many together to get a clear picture. First of all, there was a big problem—Ed.

My first reaction to the problem was not to believe it. I tried the experiment and, sure enough, it didn't behave as described. The flame trailed behind the candle as it orbited the center of the turntable, just as it trails behind as you walk along with the candle.

GARETH KELLY
HEAD OF PHYSICS
PENGLAIS SCHOOL, ABERYSTWYTH, DYFED, UK

After I read this question I could be found in the kitchen with a candle on a rotating cheese-board. At a speed of approximately 60 revolutions per minute the flame simply trailed behind the candle, showing no tendency to move out or in. I repeated the experiment later in the day on a phonograph turntable at 78 rpm, with the same result. Am I missing something?

JOHN ASHTON

Yes, Mr. Kelly and Mr. Ashton, you are missing something, although we commend your industry and integrity. So first of all . . .—Ed.

To see this effect, the candle must be effectively enclosed, otherwise it streams backward. So, candle in jam jar, jam jar on edge of turntable.

DAVID MAY
PHYSICS TEACHER
HIND LEYS COMMUNITY COLLEGE
SHEPSHED, LEICESTERSHIRE, UK

The reason the candle flame points inward is that the rotating table sets up a weak centrifuge.

DAVID BLAKE

As the air in the jam jar is being spun in a centrifuge, the denser air moves out with predictable consequences—Ed.

The candle flame bends toward the inside of the turntable for the same reason that flames move up rather than down. The heated gas of the flame is less dense than the cooler surrounding air, and the denser surrounding air moves out, forcing the candle flame in.

If I were to get really picky, I would argue that the less dense candle flame is accelerated more by the same centripetal force. Newton's law says that for the same force, the product of mass and acceleration is the same. So if the mass is smaller the acceleration must be more. At school level, it's simpler to think that the force has more effect on the denser air.

SUE ANN BOWLING
UNIVERSITY OF ALASKA
FAIRBANKS, ALASKA

You can also think in terms of reference frames or do the math—Ed.

Understanding why the candle flame points inward is made easier by considering a similar problem in a linear reference frame. Imagine you are driving in your car and in it is a helium balloon held by a string. You brake hard. What happens to the balloon? While you slam forward against the seat belt, the balloon goes toward the back of the car. This is because the air in the car has inertia and continues forward just as you do, and the balloon reacts by floating toward the lowest-pressure, lowest-density portion of the air mass at the back of the car.

Similarly, the candle flame is buoyant, its shape resulting from a complex interaction between the hot burning wax at the wick and the heating of the surrounding air. So, the flame also floats in the direction of lowest pressure—toward the axis of rotation. To complete the comparison, the candle, like the car, is accelerated with respect to the air surrounding the flame, so the air is moving radially outward relative to the candle. The flame reacts by floating inward.

TOM TRULL
UNIVERSITY OF TASMANIA, AUSTRALIA

The air in a closed container would displace the less dense gases in the flame toward the center of rotation under the centripetal force field. The flame will make an angle $\arctan(a/g)$ with the vertical (where a is the centripetal acceleration).

This effect is demonstrated by a helium-filled balloon in a car. The balloon leans forward under acceleration, backward when braking, and toward the inside of bends. The same formula applies. For a car rounding a curve of 65 feet radius at 30 miles per hour the lean should be about 44°.

NEIL HENRIKSON
RECTOR, JAMES YOUNG HIGH SCHOOL
EDINBURGH, UK

And a simpler demonstration of the same effect—Ed.

If you place a spirit level on the turntable pointing away from the center like a bicycle wheel spoke, and rotate, the bubble quickly moves inward. The more massive spirit has pushed the lighter bubble there.

COLIN SIDDONS

⚙ Ten-Bob Swerver

I am well aware (having played many ball sports) of the Magnus effect, which causes a ball that is spinning clockwise (when viewed from above) to swerve toward the right. Similarly, a ball struck with backspin will travel with a long, floating flight. These effects can be seen with leather soccer balls, tennis balls, and table tennis balls. However, when you apply spin to one of those plastic soccer balls sold at gas stations and on beaches, the opposite is observed: clockwise spin produces right-to-left swerve, and backspin produces a viciously dipping shot. These balls are really only larger versions of table tennis balls, and similarly devoid of dimples and surface markings, so why should their responses to spin be opposite?

RICHARD BRIDGEWATER

This phenomenon was dealt with in some detail in a feature called "The Seamy Side of Swing Bowling," which appeared on page 21 of New Scientist *on August 21, 1993, and is best explained in terms of "boundary-layer separation."*

When a ball travels through the air its surface is covered by a thin coating of air that is dragged along with it. Beyond this lies undisturbed air. Between the dragged air and undisturbed air lies a thin boundary layer. At the front of the ball, this layer moves slowly. But as it travels around the ball, it speeds up and exerts less pressure (as dictated by Bernoulli's law, which states that the faster a fluid flows, the less pressure it exerts).

At some point, the boundary layer separates from the ball's surface. If the ball is smooth and not spinning, this happens at the same point all around the ball. But if the ball is spinning, the boundary layer separates asymmetrically, so the boundary layer covers a larger area on one side than on the other. The result is a larger region of low pressure on one side of the ball than the other, which pushes the ball to one side.

In a conventional swing (produced by the Magnus-Robins effect), the spin of the ball carries a very thin layer of air along with it. This pushes the point of boundary-layer separation toward the back on the side of the ball where the spin is moving in the same direction as the surrounding airstream and toward the front on the side that is moving against the airstream. The result is lower pressure on the side where the boundary layer has become extended, which causes the ball to swing in that direction. That's why a clockwise spin causes the ball to move from left to right. (Another way of describing what happens is to say that the shift in the point of boundary-layer separation pushes the flow lines of the air around the ball—the ball's wake—to one side, so that the ball swerves to the other.)

All this assumes that the flow in the boundary layer is laminar, with smooth tiers of air on top of each other. In practice, part of the airflow may be turbulent, with air moving chaotically throughout the layer, and this is where reverse swing can occur. Experiments show that turbulent layers stick to the surface of the ball longer than laminar layers. So if the boundary layer is turbulent on one side and laminar on the other, the pressure will be lower on the turbulent side and the ball will swing to that side.

Under certain circumstances, turbulence can develop first on the side of the ball which is moving against the airstream, so that the boundary layer here separates later. The result is a reverse swing. Whether turbulence will develop depends on the type of ball and its speed, size, and spin, so reverse swing is seen more commonly in some sports than others (see the following answers).

Sports such as cricket, which use balls with seams, give bowlers additional opportunities to produce both swing and reverse swing through turbulence. Skillful players can bowl so that the ball spins with its stitched seam always facing at a particular angle to the oncoming air. The seam affects the airflow, making the boundary layer turbulent on only the seam side of the ball. The boundary layer thus separates later on this side of the ball and the result is a vicious swing.

Bowl fast enough and that swing can be made to reverse. At the very high speeds produced by world-class bowlers (more than 80 miles per hour), the air moves so fast that the boundary layer becomes turbulent even before it reaches the seam of the ball. In this case the seam pushes the boundary layer away, encouraging it to separate from the ball earlier on the seam side. The ball then unexpectedly swerves in the opposite direction from usual. This is the notorious ten-bob swerver.

The effect can be produced by ordinary cricketers too, if their ball is scuffed, as a rough surface allows a turbulent boundary layer to develop more easily. Deliberate scuffing is, of course, against the rules—Ed.

The reverse swerve body on a plastic soccer ball is due to boundary-layer separation. On the side of the soccer ball where the relative velocity of the air and soccer ball is larger, the flow in the boundary layer becomes turbulent. On the other side it remains laminar. The laminar boundary layer separates from the ball's surface. By contrast, the turbulent boundary layer remains in contact with the surface farther around the ball. As a result, the wake behind the ball is deflected in the direction opposite to the rotation of the ball. And it produces a force toward the side of the ball that is moving in the direction opposite to the airstream (from right to left for a ball spinning clockwise).

Experiments show that the main factor governing the direction in which a ball swerves is the ratio of the rotational speed of its surface to the ball's translational speed. The reverse swerve occurs when this ratio is small (less than 0.4), whereas the Magnus effect occurs at higher ratios. This probably explains why the faster-spinning tennis ball swings in the direction opposite to the soccer ball.

OLIVER HARLEN
UNIVERSITY OF LEEDS
WEST YORKSHIRE, UK

The swerve of a spinning ball is commonly ascribed to the Magnus effect; but more than a century before Heinrich Magnus, Benjamin Robins studied spinning cannonballs, and in 1742 he published a description of why, even on windless days, they swerved off course.

BRIAN WILKINS

Many publications do now refer to the Magnus-Robins effect. It is perhaps worth remembering that Isaac Newton commented in 1672 on how the flight of a ball was affected by spin—Ed.

◈ Red-Hot

What causes the colors that form on a clean iron or steel sur-
face after it has been heated and cooled for tempering? The
colors range from yellow when the metal is heated to about
400°F, through gold, brown, purple, blue, and finally black
when it is heated to about 1100°F. And because the oxidized
blue or purple finishes on steel mechanisms have often sur-
vived unmarked in clocks from the nineteenth century, what
is the physical nature of this transparent and very durable
colored layer?

JOHN ROWLAND

The hot furnace gases that are used for heat-treating steel
oxidize the alloying elements, such as chromium, to form a
thin surface film. These surface films interfere with visible
light waves to produce the colors that your correspondent
mentions.

The thickness of the films determines the apparent color
of the steel as it interacts with light of different wavelengths.
Thinner films, which are formed at lower temperatures,
seem yellow or gold. Thicker films make the steel appear
light blue. The thickest films seem midnight blue and finally
black.

Temper colors on clean, bare steel are actually quite fragile, and are quickly lost if rusting thickens the surface film by
depositing layers of hydrated iron oxides. Many parts of the
hundred-year-old clocks mentioned in the question owe the
durability of their temper colors to the practice of dipping
the tempered steel in sperm whale oil. The sperm oil gives a
transparent, waxy protective covering to the oxide films,
preserving their colors for posterity. Widespread use of this
technique has had the obvious disadvantage of producing a
serious shortage of sperm whales.

DALE MCINTYRE

◈ Air Space

We have tried the experiment demonstrated by science teachers in which a candle standing in water is covered by an upturned glass. The candle goes out and the water level rises in the glass.

We are taught that the water level rises because oxygen is being consumed by the burning candle. However, if we have four candles burning under the glass instead of one, the water level rises much more. Why?

EMMA, REBECCA, AND ANDREW FIST

Emma, Rebecca, and Andrew's questioning of the seemingly well-understood candle experiment demonstrates how young and inquisitive minds are able to demolish false explanations propagated through school physics over the decades.

The consumption of oxygen may well contribute to the rising water level to a certain extent, because a given mole volume of oxygen will burn the wax's carbon into roughly the same mole volume of carbon dioxide and the hydrogen into two mole volumes of water vapor respectively.

The former will partly dissolve into water; the latter will almost completely condense into liquid water. This will certainly lead to a net decrease in gaseous volume.

However, this is a minor consideration: the important influence is the heat created by the burning candle(s). By the time you cover the candle(s) with an upturned glass, an increased number of candles will have increased the air temperature around themselves more than a single candle would.

As soon as the candle(s) go(es) out, the surrounding air contracts as it cools and the ratio of contraction is directly proportional to the initial average temperature of the air volume under the glass. So more candles lead to more heat, a higher temperature, and a higher water level upon cooling down.

All this tells us that we should never believe science teachers without asking a few pertinent questions first.

LEOPOLD FLATIN

Congratulations to the children who experimentally disproved the common textbook misconception about the candle, the upturned jam jar, the dish of water, and the alleged removal of all oxygen from the jar.

By observing that four burning candles cause the water to rise significantly higher up the jar, they have shown that the principal cause of this effect is the heat from the candles, causing the air in the jar to expand. They will also have noticed that the expanded air makes a "glug, glug" sound as it escapes around the rim. There is a short pause after the candles go out, and only then does the water level rise as the remaining air cools and contracts again.

A candle flame goes out after only a small percentage of the available oxygen has been used up. So it is wrong to claim that this experiment demonstrates the proportion of oxygen in the air in some quantitative way.

IAN RUSSELL
INTERACTIVE SCIENCE LIMITED
HIGH PEAK, DERBYSHIRE, UK

The effect is partly caused by the thickness of the three extra candles. You get the same effect using single candles of different thickness. The thicker the candle, the higher the water will rise.

The water drawn in is squashed into the space between the candles and the glass. The narrower this space, the higher the water will rise.

PETER MACGREGOR

⬙ Deflation Policy

Why do helium balloons deflate so quickly? When my children bring balloons home from parties, the ones that are filled with helium are often small and wizened by the following morning. I realize that some of the size reduction is caused by deflation, but something else must be at work, because standard air-filled balloons stay inflated for much longer.

JOHN STORR

Helium gas is not only very light but monatomic—its particles are all made of a single atom. As a result, helium is made up of the smallest gaseous particles possible. The atoms are only 0.1 nanometer in diameter, and are quite capable of diffusing through metal films. Because it so readily diffuses through small pores, helium is used to help test for leaks in industrial and laboratory vacuum systems. Nitrogen and oxygen molecules have a much larger diameter than helium atoms, so they are much less capable of diffusing through the balloon walls. It's like the difference between trying to get sand and small pebbles to pass through a sieve—the sand goes through much more easily because it's made from smaller particles.

The second factor which helps to increase losses by diffusion is that balloons are made from viscoelastic materials whose structure is a tangled mass of polymer strands—a bit like a plate of spaghetti. The polymer strands cannot pack closely together, and have channels through which the helium can diffuse, so even at low pressure the helium will diffuse out through the walls. When the balloon is inflated, the polymer stretches, so the balloon walls become thinner (the helium has a shorter distance to diffuse out), the molecular structure becomes slightly more open (making diffusion much easier), and the increased pressure provides a driving force for the diffusion. These are the reasons why deflation is very rapid to begin with, but then gradually slows down as the balloon gets smaller.

Commercial helium balloons are made from nonporous inelastic materials and are coated to reduce the losses even further, although even they still lose a significant percentage of helium per day, certainly enough to disappoint children (and grown-ups) the morning after they buy a balloon.

GAVIN WHITAKER

The helium atom is very small and very light. It is able to diffuse through the thin, stretched rubber of the balloon quite easily, finding its way through atomic-sized pores. Air molecules (oxygen and nitrogen mainly) are larger and heavier and diffuse through much more slowly. In addition to the increased pressure inside the balloon which pushes helium out through the sides, there is another factor that increases helium flow outward.

Because there is almost no helium in the air, far more helium atoms are hitting the inside of the balloon than the outside, and there is a net flow outward. However, you will notice that the balloon does not completely deflate. This is because some air moves in as, conversely, more air molecules hit the outside than the inside.

This leads to a truly bizarre effect if the balloon is filled with the gas sulfur hexafluoride, which has large, very heavy molecules that hardly diffuse through the rubber at all, and so cannot get out. But once again, as in the helium example, there are more air molecules outside than in, so air diffuses inward and the balloon slowly increases in size.

HARVEY RUTT
DEPARTMENT OF ELECTRONICS AND COMPUTER SCIENCE
UNIVERSITY OF SOUTHAMPTON, UK

⚙ Shafted

If you find yourself in a free-falling elevator, is there any action that you can take to reduce the effect of the collision? Would jumping just before you hit the bottom of the elevator shaft help?

NIGEL OSBORN

First of all, Hollywood clichés notwithstanding, it's almost impossible for an elevator to fall down its shaft, thanks to Elisha Otis's nineteenth-century patent acceleration-sensitive safety brake. The instant a car starts to fall, multiple spring-loaded arms pop up and wedge it in its shaft.

As for improving your survival prospects, probably the best thing you could do is lie faceup with your back on the floor and your hands under your head to minimize the impact, although this would be difficult to do if you're in free fall.

Jumping just before impact would merely delay your own impact by a few milliseconds. Besides, how would you know when to do it? If you jumped a moment too soon, first you'd bang your head on the ceiling, then you'd be slammed to the floor when the lift hit the bottom.

Even if you could time your jump precisely, to do any good you'd have to exert the same force required to jump to the height from which the elevator fell (for example, if the elevator fell 300 feet, only people capable of jumping 300 feet in the air could save themselves that way). If they could do that they probably wouldn't need the elevator.

KEITH WALTERS

If you jumped a moment before hitting the bottom, giving yourself an initial upward speed, relative to the elevator, equal to that of its downward velocity, you would head swiftly toward the roof of the elevator compartment. There would be problems with jumping, as you would probably be weightless, but with handles to let you pull against the floor it ought to be possible.

Fortunately, just before you hit it, the roof would suddenly accelerate very quickly away from you (assuming the elevator kept its shape after impact!) until it had the same (relative) upward velocity as you. Also, the floor would do the same, but toward you. You could then land lightly from a few inches off the floor and walk out of the elevator onto the ground floor, which would be traveling upward at the same (relative) velocity.

However, there are one or two problems with this. To achieve such a velocity, you would have to be capable of jumping the height from which the lift dropped. And even if you could do this, one surmises that the acceleration produced in order to jump would be comparable to that experienced on hitting the bottom.

Even so, by similar reasoning, you would suppose that even a small jump would lessen the impact.

ALEX WILSON

I can see three ways of increasing your chances of survival, although only slightly. The first has already been mentioned— jumping as vigorously as possible before you land in order to cancel out some of the upthrust. The second is to get any soft objects you have with you, your clothes for example, and place them underneath yourself prior to impact. This would increase the deceleration time of the collision, and slightly reduce the amount of damage done. If you aren't bothered about your legs, I suppose you could try standing up, and have them act as "crumple zones," although this could be fairly messy. The third is hardly worth mentioning. You could try to spread yourself out as much as possible while holding on, in order to increase the elevator's surface area. This should decrease its terminal velocity by some indiscernible amount.

DAVID FOALE

In Black and White

While working at a factory that produces carbon powder, I noticed I had made a large black thumbprint on one of my sandwiches. This set me wondering why bread and for that matter potatoes, rice, and sugar, which are mostly carbon, are not black.

DOUGLAS THOMPSON

The best way to explain is with an example. Sodium reacts violently on contact with water, and chlorine is a highly toxic greenish-yellow gas. However, sodium chloride, the compound which contains these two elements, is harmless common salt, showing that the properties of an element are very different from the properties of that element's compounds.

The black powder used to produce a photocopy is finely ground carbon in its elemental form. The particles are extremely small and arranged at random. Any light which falls on them is absorbed and not reemitted, so the powder looks black. The sandwich certainly contains carbon but not in its elemental form. Here, it is combined with oxygen and hydrogen as carbohydrates. These compounds have their own properties, which are nothing like the properties of their constituent elements. The slices of bread emit light of many wavelengths reasonably well, so when we look at bread in daylight, it appears white.

Richard Honey

Carbon is normally found as an amorphous solid, which lacks a definite crystalline structure. Because of this, and because of the position of certain electrons in the outer orbits of the carbon atom, light is absorbed and not reemitted. As a result, the carbon atoms in graphite, soot, and carbon black appear black.

Diamond, which is also carbon, is normally clear, because its crystalline structure alters the electrons and their positions to create a colorless crystal. Diamonds can be colored if other atoms, usually metals, are present and alter the electron bonds to create blue, yellow, pink, and green versions.

H. William Barnes

Carbon, as present in foodstuffs such as bread and potatoes, is in hydrated form—the carbon has been chemically bonded with water and so does not appear black. To get the black carbon back you need to remove the water, usually by heating. This is why burned toast is black.

Sugar is also carbon and water. But add concentrated sulfuric acid and you'll see black carbon appear as the acid efficiently sucks out the water.

Duncan Hogg

Index